LITTLE JOE

THE STORY OF

JOSEF FILIPOVIC

SURVIVING IN A WORLD AT WAR
WHILE FIGHTING ANOTHER AT HOME

BILL WHEATLEY

"Little Joe," by Bill Wheatley. ISBN: 978-1-62137-344-5 (softcover); 978-1-62137-345-2 (eBook).

Library of Congress Control number: 2013913336

Published 2013 by Virtualbookworm.com Publishing Inc., P.O. Box 9949, College Station, TX 77842, US.

Manufactured in the United States of America.

for Heidi and Stefan

A NOTE ABOUT NAMES:

At birth his name was Josef Filipovic, pronounced *YO-seff fi-LIP-o-vich* in Serbo-Croatian. His nickname as a boy was Josko, phonetically *YOSH-ko*. When he moved from Yugoslavia to Holland, the spelling of his surname was changed to Filipowic, also retained when he came to America. In recent years, he preferred to use the original spelling and pronunciation.

The names of the people in the story have been changed, except for Joe's family. The other exception is the Bingham family who played a historically significant role in the reconstruction of postwar Germany, as well as in Joe's life.

Special thanks to Murdo Macphail for translating and decoding my transcription of the German words and phrases in Joe's story, and to Ted Tumelaire and Ralph Kaefer for similar help with the Dutch ones.

Any mistakes are mine.

Bill Wheatley

CONTENTS:

Introduction
Boston, 1969

"You want to meet this guy coming from New York?"

"Sure," I said. "I guess so."

I did and didn't want to meet the guy coming from New York. I was working as a film editor at a public television station in Boston and the new guy, Joe Filipowic, was an editor based in New York who was being imported to save a long documentary which was defying the younger, less experienced editors who had been working on it for several months. The airdate was approaching and they still hadn't pulled it together. I wasn't one of the editors who had been humbled by this show but I still felt threatened by Joe's invasion. He was a tough pro from New York who had worked with some of the big names in network television news and public affairs, and we were ambitious kids—filmmakers, not real professionals. At the same time we wanted to be pros ourselves; we had to meet deadlines as firm as an executioner's and were proud of how hard and long we could work. So it jolted us to hear that the week before Joe arrived he had put in more than one hundred hours and that he often did that several weeks in a row to finish a film for broadcast. But his final attribute, the one we couldn't compete against, was something that made him potent and mysterious beyond the skills of editing films: his background. He was European, a Yugoslav who had somehow survived the Germans during World War Two and made it to America. His unknown history seemed summarized in the rumor that he spoke seven languages. In self-defense, I was determined not to be impressed by this new man.

Joe arrived on a Sunday evening. I was hanging around, working on a different film, when one of the other editors invited me to come down to the news office where the others were waiting to meet Joe when he got in from the airport with the producer. There were only a couple of desk lamps lighting the large, low-ceilinged office, and in one corner a television monitor was turned on, coincidentally showing the broadcast of the program Joe had just finished editing in New York. As if to make way for the visitor we were spread out around the perimeter of the dimmed room watching the television when he walked in. Joe looked so close to my preconception of him that that itself was a surprise. He was just under medium height, not broad but thick, and he walked tilting forward, setting each foot down emphatically which made his torso swing from side to side. His most impressive feature was his head. It was large for his body but because his physique had such an attitude of compressed strength his oversized head made his body seem even more powerful instead of insufficient to carry it. He wore his dense black graying hair in a crewcut and had distinctly Slavic features; a narrow nose which thrust down like a slash and high wide cheekbones jutting down to his chin. Like his motions and gestures, his voice was emphatic, and accented.

Joe and I had little to do with each other at first. I was immersed in finishing the film I was editing and he dove into the one he had come to save. Fueled by coffee and cigarettes, he worked the long hours we had heard about and instead of being harassed by the looming airdate, he seemed as happy as a man can be. He supervised the editing, instructing his two assistants, and dazzled us all with the way his muscular hands flew over the splicer as he whipped the strands of celluloid and magnetic film into new sequences. He presented the producer with alternatives, and despite his own strong opinions, the arguments were friendly. The atmosphere which had been tense and nearly desperate became merely urgent, and collegial, although Joe was in charge.

Except to peek in once in a while, I was too busy to watch him work, but I heard from the other editors about his quick, decisive techniques and formidable energy. The few times I talked to him casually or asked a technical question, he was more than friendly; he explained his answers like a conscientious teacher. The one event involving the two of us during that time, however, was different.

One night, Joe and I were both working late trying to finish our films. There was only one editing machine available and we both wanted to use it so an arrangement was made to share it during alternate blocks of time through the night and into the next morning. In the midst of my work, I left the editing room to get a piece of film from another room and came back to find Joe in my chair, threading up his film with mine pulled off and set to the side.

"Hey, what's going on?" I protested. "You're not on for another hour!"

"Yes, we are. Two o'clock," Joe said, starting his film.

We started arguing and by the time the two producers came to see what was going on, Joe and I were occupying the same space in the doorway, each trying to block the other out. For a moment it seemed as if it was going to become a tag-team wrestling match, but my producer reminded everyone of the agreement and the other producer urged Joe to relent. The struggle ended and we had another hour. Joe had made an honest mistake, but it was one his seniority inclined him toward. He was the veteran editor, I the novice, he resented the fact that the station was so under equipped, it was an insult to have to work in conditions like this. I sensed all that in his attitude, and respected it, but felt my own urgency to get my film done.

The next day the story had gotten around and there was some chuckling over it. Everyone got a little desperate trying to finish their films in time to meet airdates, but starting a fight with Joe was crazy—his manner was considerate, almost formal at times, but he radiated such an intense force it seemed incredible anyone would risk setting him off. I felt embarrassed and foolish, and avoided his editing room. Then I ran into him by accident and before I could say anything Joe apologized, taking all the blame for the misunderstanding. We went to have a cup of coffee together, laughing about the whole thing. A few days later Joe's producer stopped me in the hall at the station and said, "I thought you and Joe were enemies. I just heard you're sharing an apartment." It was true. On weekends when he wasn't working, Joe flew home to see his wife and children in New York but he was spending so much time in Boston he needed a place to stay during the week. He had found an apartment and wanted someone to share the cost, and by coincidence, I'd just lost my digs. All he needed was the small rear room he slept in and when I moved in he insisted I put my bed in the large empty living room.

It was a noisy place. At night I was wakened by the massive trucks, eighteen-wheelers that rumbled down the street and shook the whole neighborhood when they stopped for the traffic light in front of the building. I was still learning to sleep through these monsters when one night I was wakened by a loud banging in the kitchen. I wondered if something was wrong with Joe but didn't want to to wake up completely so I ignored it. A moment later there was another crash. I listened to the silence that followed, trying to decide whether to investigate. The third crash got me up. I pulled on my pants and went into the kitchen. I was tired and annoyed but the sight was funny. Joe, in his usual Brooks Brothers button-down shirt with the cuffs rolled up twice under the sleeve military style and with his tie pulled down, was bent over the broiler. In his hands he held the black, encrusted pan and on the pan was an inch-and-a-half thick frozen steak larger across than Joe's two hands. The racket was caused by Joe trying to force the pan to slide into the grooves under the broiler. I watched him for a moment, his strong back twisting as he tried to wrestle the pan into the grooves. When he paused after another try, I said quietly, "Joe..."

He looked around, startled.

"Uh...sorry, Bill. I forgot you were here."

"That's okay," I said. He looked like he was drunk. "Don't worry about it."

"I didn't mean to wake you up. I'm trying to get this..."

He bent down again and I said, "Here, let me try it."

Joe let me take the pan and after some jockeying I got it to slide into the grooves, then turned the knob till the burner lit with a small pop.

"Thanks," Joe said, his tongue even heavier than usual on the 'th'. "How about a beer?" He opened the refrigerator which was nearly empty except for beer and stumbled as he turned around to hand me one.

"Uh, okay," I said. "What the hell." I was wide awake now, and I also didn't want to leave him alone. I hadn't seen him like this before and he looked like he needed company.

"Did you work all night?" I asked, taking the beer.

"Yeah. Then we went to this after-hours joint, someplace downtown..." Joe chuckled.

We sat down at the metal-topped kitchen table, waiting for his steak to cook. Talking with Joe was always enjoyable for me—even at three in the morning—because he had a good sense of humor and always saw the bottom line of things. He also had unpredictable, often original opinions and sometimes asserted himself in ways you couldn't contradict. I assumed these undeniable opinions came from his early experience and although I was curious about his past, I never asked any questions. I sensed it was something he couldn't talk about in ordinary conversation.

While we were talking, Joe got up and turned his steak over and when it was done he put the brown, naked meat on a plate, offered me some which I declined, then ate the whole barely cooked mass like a carnivore. It went so fast I hardly noticed him eating. When he was finished he got us both another beer, sat back down in the creaky chair and lit a cigarette, looking tired now instead of drunk.

"You shouldn't work so hard, Joe," I said. "You're killing yourself."

Joe shook his head resolutely as if I were suggesting something impossible.

"I know you've got to do it sometimes just to get the job done and you want the extra hours—overtime and double time and all that—but you can't keep pushing yourself like that forever."

The pressure of television was exciting and I'd learned to work as long as Joe when I had to, but I was younger than him and I knew what it took out of me.

Joe was still shaking his head.

"Let me tell you something, Bill. I work like this for one reason. That is my children. Everything I do is for them. That's what all this is for, the nights like tonight when I've been working for..."

He looked at his wristwatch and continued staring at it, as if the numbers and hands had no context. He'd obliterated the normal sense of time, there was no night or day, no seasons, the year was divided into shows he'd worked on, marked off by the paychecks he'd received.

"I know, Joe, but—"

"You see, Bill," he interrupted. "This doesn't matter. It doesn't matter how many hours I work or whether I...I...this is my job. This is what I can do. This is my shot and I have to..."

He was so tired, his feelings so intense, he could only repeat himself.

"I know, I'm just saying you don't want to kill yourself, especially for their sake...."

He was shaking his head.

"Bill...the way I can respond to that is to tell you a story."

He sipped from his glass, poured more beer down the side from the can, and set both glass and can down on the table. Then he leaned forward on his elbows, inhaled the last of the cigarette, and said:

"This is now winter...we are unloading these logs...it's cold, snowing...they had brought in these railway cars loaded with logs, stacked up high, like this."

He raised his hand up over his head. The way he pronounced the word logs sounded like 'lawks.'

"They were using these for construction in the coal mines."

He paused to mash out his cigarette and I asked:

"This was...during the war?"

He nodded. "In the camp."

"The...?"

"The labor camp."

"In Germany?"

He nodded again.

"We had no gloves, no coats, just what you wore, your shirt and wooden shoes that you packed with paper stuffed in there, whatever you could find. The hands got cold, stiff, so you could hardly hold the logs. They were heavy. They rolled them off the car and you took it on your shoulder, then you had to walk over and throw it on the pile. The guards were all around. Old German soldiers with rifles, overcoats, boots, you know, they were warm."

Joe lit a cigarette.

"You had to carry the log down a bank, it was muddy, the snow was melting there where it fell, and all the time the Germans are shouting at you, shouting—wherever there are Germans there is noise—*'Schnell, schnell! Mach schnell! Mach schnell!'* Anyway, one of the guys, when the log comes off the railway car he takes it on his shoulder and as he is coming down the bank he swings the end around and hits the guard with it."

Joe stopped, staring at the table, as if that was the end of the story.

"You mean he hit the guard with the log?" I asked.

"Yeah. Hit him with it."

"What happened to the guard?"

"Oh, he is dead. Immediately."

"What about the guy, the guy who did it?"

"Well, now everything is chaos—shouting, cursing, guards all over the place, and an officer comes running up. 'What happened?' So. They line us all up. 'Who did it?' No one says anything. The officer is furious. 'Who did this?' He shouts and threatens."

Joe dragged on his cigarette.

"None of the other guards saw who did it?"

"No. Nobody saw it. There were maybe fifty, sixty guys there, working. It was all very fast. The guy hit him, he was dead. There were several guys there that could have done it. Guys who had an attitude that, they're not taking this shit."

Joe paused.

"So the officer, he is SS, he says, 'All right. I give you ten seconds.' No one speaks. We stand there in the cold, the mud, snow is falling. Then he says, 'Count off! *Eins, zwei, drei! Eins, zwei, drei! Eins, zwei, drei!*'"

Joe's voice was suddenly loud, barking out the commands in German which sounded more natural on his tongue than English.

"All the way down the line. Then: 'All right—third man, out. *Raus!* The third man steps out and the guard comes down the line and shoots them in the back of the head."

Joe's thick hand was around his glass of beer; he picked up the can and poured out the dregs.

"What?" I said quietly, surprised by the abrupt end of the story. "They shot a third of the men?"

"Every third one."

"And you were..."

"Number two. *Zwei.*"

"So the guy next to you, he got shot."

"In the back of the head. You saw the shit splash."

"The 'shit'—"

"The guy's brains."

He inhaled his cigarette which was out, burned down to the filter. He dropped it in the empty beer can, lit another, then got up to get another beer from the refrigerator. He put another one down in front of me too.

"Jeez, Joe..." I said, shaking my head, watching him pour beer into his tilted glass. The cigarette jutted from his fingers, the blue eyes behind his glasses watched the beer fill the glass.

"What happened after they...shot everyone?"

"Then we put the bodies in bags and loaded them on another railroad car and started unloading logs again."

I sat there, still not knowing what to say, and when I finally popped open the tab on the can it sounded rude, as if it was mocking the silence.

"How old were you when this happened?" I finally asked.

When Joe didn't answer I thought he was so lost in memory he hadn't heard, or just didn't want to say anything more.

Then he murmured, "Fifteen."

"Fifteen," I repeated to be sure I'd heard right. I'd pictured someone older, not just a kid, even though I knew he could not have been an adult yet.

"How did it...I mean, you were so young—"

The sound of Joe inhaling deeply cut me off. Then he spoke quickly, getting it out of him.

"I felt like a whore of a man."

I couldn't quite follow.

"A whore?"

He didn't answer.

"You mean because the guy next to you...and you were still..."

Joe was staring at his glass. It seemed wrong to say anything, as if any response but silence was wrong.

We sat for a while.

"For the moment it happens, there is a relief," Joe said thickly. "Then, after it happens, there is no relief."

There was another long silence. I wanted to say something to change things, to shift his burden.

"Have you ever told your kids about...what happened?"

He said, "No."

"You really should, Joe, I mean sometime at least, when they're older anyway."

He was shaking his head, vehemently this time.

"All this is—this is not—this is why I work so hard, Bill, so they don't have to..."

He faltered, losing the words for what he felt.

"You should tell them just so they know who you are and what you've lived through," I went on, knowing it wasn't my business, but thinking of that familiar figure, the maniacally hardworking immigrant parent estranged from his own American kids who can't understand him.

Joe was still shaking his head but was too tired to answer. He had worked his eighty hour week, had gotten drunk, eaten his steak, and told a traumatic story. Now he was spent.

We sat a while in silence, feeling the cold in the room.

Joe yawned and got up. I stood up too. It was time to get some sleep.

The window into Joe's past shut after that night and didn't open again until a couple years later when I ran into him at the restaurant across from the film studio. He had been in Geneva, Switzerland, working as the supervising editor on a series of documentaries for public television on the Arabs and Israelis. It was a prestigious job, but in the meantime other things had happened too. Joe's wife had filed for divorce and I knew without hearing any details how much that would tear up his life. I also assumed without knowing that Joe would not be an easy man to be married to. To be intimate with him would mean getting close in some ways to experiences like the one he told me about that early, boozy morning in the apartment. Now, having a beer in the Dog & Pony, we glided over the personal stuff and talked mostly about work, the film business, politics. The bar had thinned out as people finished their after work drink and I told Joe I had to get home.

"Have another," he said, waving to the bartender.

We hadn't seen each other in a long time and Joe said, as if to keep me longer, "Did I ever tell you about the time I escaped from the labor camp?"

"No," I said, surprised he would suddenly bring up the subject, as if resuming the story begun several years before. "You escaped? I thought..." I didn't know what I thought except that it didn't seem possible to escape from such a place. "How did it happen?"

"I just walked out."

"Walked out? You mean you could just—how'd you get there in the first place?"

"Well, to describe that, I have to go back." Joe filled his glass and lit a cigarette. "This begins, really, on May 10, 1940, the day the Germans invaded Holland...."

Joe talked for the next several hours and by the time we said goodnight, he had escaped from the labor camp, crossed the German border under Allied bombardment, and been liberated by the American army in Holland.

The next time we met, I brought a tape recorder.

When we began I told Joe if we wrote a book about his early years he would have to tell everything—not just the stories that were exciting, even funny at times, like his escape—but also the experiences that were not as easy to recall. I had in mind the execution in the labor camp he'd described so unexpectedly when he was exhausted and drunk. He agreed, and on a beautiful autumn Sunday afternoon we sat down with a tape recorder and a bottle of Beaujolais Joe had bought for the occasion. Then, instead of beginning his story, nothing happened. He was unable to speak. I didn't think the problem was having the recorder sitting in front of him, because Joe had filmed many people and was used to the process, even if he was usually on the other side. Whatever it was, something blocked him from speaking, and we just sat, silent for more than an hour, avoiding each other's eyes, while the tape machine recorded our silence. I tried gently prompting him several times and he shook his head, apologized, and said to bear with him. When he was finally able to speak, starting, stopping, starting again, clearing his throat, almost groaning, his voice was strangled with emotion.

"My father...*hated* me. He despised me..." Joe began. With great difficulty he tried to describe his relationship with the most important person in his early life. Over the next weeks and months, the story of those years came out, often with his typical energy and enthusiasm, at other times struggling against his reluctance to relive the events. I had assumed his story was about surviving the Nazis, and the war. It was, but he'd also had to survive an enemy even more threatening—his own father.

CHAPTER 1
HEERLEN, HOLLAND; MAY 10, 1940

I was asleep in my bed before dawn when all of a sudden I woke up—guns were firing, shells blasting. I jumped up and got dressed, ran downstairs and out to the street. The sky was just getting light and I saw an airplane, a single plane coming from the east, from Germany, flying right toward me. It swooped over my head, so low it almost flew between our house and the neighbor's, but it didn't strafe or drop anything. Then it pulled up, roared off over Holland and disappeared. Guns were still booming in the distance, not loud but I could feel the explosions shaking the ground, and now some of the neighbors were coming out, half-dressed, looking around, asking each other what was going on, had the war started.

"Josko!" someone shouted. "What's happening?"

It was Wil, who lived in the other side of our house, and had just come out too. He was looking toward Germany where all the noise was coming from.

"I don't know," I said, "but a German plane just flew right over the house!"

Wil didn't look happy. He was German but his wife was Dutch and he had no use for Hitler.

Frankstraat—our street—was a small housing development just outside the city of Heerlen in the southern part of Holland. It was only eight kilometers from the border with Germany and ever since the previous September when Hitler invaded Poland, everyone had been talking about war. England and France had already declared war on Germany but for months, the whole winter, nothing happened—not till now.

Wil and I were still standing in the street when a man came running out of his house shouting. "The Germans have invaded! It's on the radio!" He was excited because he was German himself. There were all kinds of people living in Frankstraat—Germans, Poles, Hungarians, Yugoslavs—all the guestworkers and their families who had come to Holland for a job in the coal mines, including a few Dutch who worked there too.

I wanted to see some soldiers, some action, so I ran down Frankstraat to the main road and headed toward Heerlen. The center of town was about five kilometers away but before I got there I ran into a column of German soldiers marching along, pounding their boots on the road and singing, "*Auf der Heide blüht eine kleine Roselein und sie heisst Erika.*" They looked very cocky and sure of themselves in all their battle gear and they ignored me completely, a little twelve year old kid standing by the side of the road wearing short pants with a rip up the side and shoes made from old rubber tires. I watched the soldiers march off but there was really nothing to see, they were there, that's all, and when I went to school, assuming everyone would be there, *Mijnheer* Van Dijk was sitting alone at his desk.

"The *Moffen* are here, Josko," the teacher said. *Moffen* was a nickname we had for the Germans. I don't know what it meant or where it came from but it wasn't meant kindly.

Mijnheer Van Dijk and I did a science lesson on the expansion of metals under heat. He lit the Bunsen burner and we set up the experiment but every few minutes the bombers flew over and it was so loud he had to stop talking. Finally, he said, "Josko, go get my bicycle," and he rode off to go find his brother who was in the Dutch army in Valkenburg, not far away.

When I went home my father was there too. It was Friday and they'd shut down the mines for the weekend. Wil's wife had come over and she was sitting in the kitchen with Mama, both of them crying while they listened to the news reports on *Radio*

Hilversum. Mama had lived through the First World War in Yugoslavia and seen her father, Big George, lose most of his farm because of the war. "It's just like before, we're going to lose everything," she kept saying. "The English will have to stop the Germans." But the Germans were bombing the hell out of Rotterdam and although the Dutch army tried to make a stand here and there, they were really already conquered. The following Tuesday, Holland surrendered. I was sorry for my Dutch friends like Scheng and Leo because they felt so angry and humiliated, but other than that it didn't seem to affect my life very much.

Two years later, in the spring, I finished school. I was fourteen and unlike the nice Dutch boys from good families, I wasn't going on to the high school, the *Gymnasium*. I worked at a farm a few days a week and at the end of the summer I got a notice from the *Deutsches Haus,* the German military administration, telling me to report for work. There was nothing threatening in the letter, just that you had no choice. Many people had been called. You'd be trained, work in a factory or some place in Germany, and be paid. Nevertheless, I wasn't going to go. I hated the Nazis—the sound of Hitler's voice on the radio, shouting about the enemies of the *Reich,* especially the "bandit" Tito in the mountains of Yugoslavia fighting the "heroic" German soldiers. The Germans had invaded my homeland and to me they were the enemy. There was no way I was going to show up to work for them.

There was a Hungarian guy a couple years older than me we called the "*Dolle*"—it meant "crazy" in Dutch—and he was a big, clumsy kid who used to run up and down the street shouting, "Hi-yo, Silver!" like the Lone Ranger. He'd already gotten a notice from the *Deutsches Haus* and instead of going he disappeared. I'd heard that some people were hiding out in the water tower across the road from Frankstraat. It was out in an orchard no more than a kilometer away, and it was no longer used as a reservoir. I also heard that people sometimes brought them food, so early on the morning I was supposed to report I walked down Frankstraat, turned left on the road toward Heerlen, and then cut around the Dutch "Alps", the steep piles of stone tailings from one of the old mines, and walked across the field to the tower. When I climbed up and looked in, there was the *Dolle* and another fifteen or twenty guys hiding out too. All of them were older than me and the *Dolle* was the only one I knew, but he told them I was okay and I climbed in.

No one showed up with food and all we had to eat was the apples we got from the orchard around the tower so it wasn't long before I had diarrhea like everyone else. One night the *Dolle* decided to sneak back to his house for some food. He got caught and gave us away, and the Germans drove a truck out in the field. "Climb down, you're surrounded!" they shouted and rounded us up. They interrogated us at the *Deutsches Haus*—the *Dolle* was already there looking rather sheepish—then they put us back on the truck for another ride. This one was longer, and it was dark so we couldn't see anything, but when we finally drove through a gate past a double barbed wire fence and towers with spotlights and guards, I knew what it was.

The labor camps, the *Arbeitslager*, were built to hold the prisoners who were forced to work for the *Reich*. The camp I was taken to supplied labor for a coal mine and it was a huge complex, right in the center of a small town called Kohlscheid, in Germany. The top of the shaft looked like a gigantic erector set about six stories high with a building around it, and there was a big yard around the building with piles of logs they cut up and used for shoring in the mine. There were stables for the horses they used to haul loads, shops to build and repair things, and offices for the German administrators. Railroad tracks ran in through a gate in the wrought iron fence surrounding the entire complex. There were thousands of us who'd been rounded up from all the neighboring countries, Holland, Belgium, Poland, men and some younger guys like me as well as women. There were also hundreds of Russian POW's. Everyone worked in the coal mine and at the various jobs throughout the complex. Besides the forced laborers, there were also the German miners, administrators, foremen and guards. The guards were old German soldiers, *die alten Kämpfer*, in their green uniforms.

I was assigned to work on top of the mine shaft where the coal came up from below. Everything else came up too—stone dug out of the mine, tools, shoring timbers—all hauled in lorries, little railroad cars that ran on tracks. When they came up full of coal a huge steel arm ejected them from the elevator and sent them down to the yard where the coal was dumped. My job was to stand at the curve just below the ejector and grab the lorries as they came shooting down out of the elevator and yank them around so they didn't jump the tracks. They shot out with a loud, earsplitting bang, and there was a constant banging and clanging of metal on metal as the steel arm hit each one from behind and shot it out of the shaft. It was a hard, noisy, boring job, ten hours a day, but I saw very quickly there were others who were worse off, a lot worse. The prisoners who worked down below in the coal mine looked bad—exhausted, filthy, beaten people with blackened faces, their eyes ringed with dust. They just looked defeated. I also heard right away that as little as we got to eat they got less.

I had worked several weeks on top of the shaft when one morning one of the *Moffen* called me and some of the other guys out before we went up to work. He led about thirty of us into the main administration building just inside the gate. We walked down the hall past offices where people were working, secretaries typing, all very civilized unlike the way we lived, and were led into a classroom. Another *Mof* wearing a coat and tie with a Nazi party emblem pinned on his lapel greeted us with a smile, and said, "*Glück auf*." We said *"Glück auf,"* and he said, "Please sit down, young men."

This was something different. Usually the only time a *Mof* said anything to us it was to shout a curse, a command, or a threat, but this guy was treating us like gentlemen. *"Glück auf"* was a coal miners' expression, a greeting everyone used, and it meant literally, "luck up," if you're lucky you'll get back up from the mine. We sat down in the chairs—just like in school everyone avoided the first row and tried to sit as far back as they could—and the German stood on the podium in front.

"Young men," he addressed us seriously but also pleasantly, "we have observed you at work and you seem to be good, conscientious workers, so you have been selected for a special training program. If you apply yourselves and complete the program successfully, you will qualify to work as *Steigers*."

Steigers were supervisors who just walked around watching what other people did, giving orders, filling out papers. An easy job.

"If you cooperate with us, we will train you for several weeks and then you will be sent back to your home communities to work. You will become an example of how the *Reich* treats those who cooperate with us."

Sitting there in a classroom instead of standing up on top of the shaft yanking lorries around, I thought, fine, anything that gets me off that job and maybe even out of this place sounds good to me. But it did seem strange that of all the thousands of people in the camp we were the ones chosen for this training program because I knew most of the guys in the room and they were all bad news, totally committed to sabotaging the *Moffen* whenever they could. There were two young Dutch guys sitting there who worked on the ejection device above me and they were always finding ways to screw things up. One day they threw a slab of wood under the wheels of a lorry full of coal just as it shot out of the elevator, so when it came down to me on the curve it jumped off the rails and crashed into another lorry coming back up on another track. It made such a mess they had to shut down the whole operation for several hours. These guys would also switch tags on the lorries so the coal was sent to the waste dump and stone was sent to the coal-bath, or they'd just tear off the tags and throw them away so the German miners got cheated out of their pay for the coal they'd dug. Anything to screw up the system. So we were hardly the kind of guys you'd select to work for the *Reich*. But we said to ourselves, if you want to retrain us and let us go, fine.

Immediately, conditions improved. We were moved to a barracks closer to Kohlscheid, we got marmalade for our bread and real coffee for breakfast, a lunch at noon—none of the other prisoners ate lunch—and several times we even got meat for supper. Meat was unheard of. Every morning we had calisthenics, then classes in German language and history, all very much the National Socialist version, and in the afternoon we worked in shops learning the various trades used in operating a coal mine. I was assigned to work in a carpentry shop where we built forms and cut the shoring timbers they used down below. We'd had three weeks of the new routine when one Friday the teacher announced that next week we would have our first class in political indoctrination. I thought we were already getting plenty of that in the history lessons. When we filed in on Monday we stood by our desks waiting for the teacher like we always did. He came in and closed the door but instead of going to his desk like he usually did he stepped briskly up to the podium, shot his arm in the air, and said, "*Heil Hitler!*" We stood there for a moment sort of confused because every other time he had greeted us "*Glück auf*" and we had greeted him the same way. So we said, "*Glück auf*," and sat down.

"Stand up!" he shouted. We stood up again. He looked at us seriously. "From now on, young men, we will no longer greet each other '*Glück auf*.' We will use the *Führer's* greeting." He paused, looking around the room at everyone, then very deliberately he raised his hand in the Nazi salute and said, "*Heil Hitler*!"

There was another pause and then we said in the usual casual way, "*Glück auf*," and started to sit down again.

"*Stehen Sie auf*!" he shouted. We stood up quickly. "You will greet me as I greet you!" Again he shot his arm up and shouted, "*Heil Hitler*!"

And again we responded, "*Glück auf*."

He was furious. He stepped down from the podium, walked over and stood in front of Jan, a tall, skinny, Dutch kid. He stared at Jan till Jan was looking at him, then he raised his arm in the salute and said in a loud, firm voice, "*Heil Hitler!*"

Jan hesitated, and then very quietly he said, "*Glück auf*."

"Heil Hitler!" the teacher screamed.

This time Jan didn't say anything. He just dropped his head and looked at the floor. The teacher went berserk.

"Heil Hitler! Heil Hitler!" he shouted over and over, and when Jan still didn't respond, he started beating him with his fists.

Jan put up his hands to protect himself but the teacher kept beating him till he fell on the floor. Then he sort of glared at us and stomped out of the room.

We stood by our desks waiting. I felt very confused by the teacher's behavior. Why had he become so belligerent? I would have done the same thing as Jan if he'd singled me out and so would any of the others. Not because we were heroes trying to make a stand against the *Reich*—we were nothing to them anyway—but because we hated the Germans and to give anyone the Nazi salute and Hitler's greeting would have been a disgrace.

When the door opened again, the teacher marched in with *Herr* Schmidt. He was the administrator of the whole complex and as a Nazi Hollywood couldn't have done better. He was tall and stout but he walked with a very firm step, and when I saw this shining, clean-faced Hun standing in front of us in his brown shirt, black boots, swastika armband, red face and completely bald shaved head, I knew we were in serious trouble. Because I had met *Herr* Schmidt once before.

I was in the carpentry shop one day when the door opened and we all stopped working and looked up to see *Herr* Schmidt standing in the doorway with four officers, two *Wehrmacht* and two *SS*. He looked around the shop as if he was trying to find someone and in a loud voice—everything he said was loud—he shouted, "Filipovic!" I raised my hand and he waved me over. "Come with me." He led me and the four officers down the hall, past the offices and display case where they had samples of all the things made from coal, and into his office. It was huge. He motioned for the officers to sit in the chairs in front of his desk and placed me to one side before he sat down in his big leather chair. Behind him was a Nazi flag and to my left, along the wall, was a bookcase. On top, lined up neatly and resting on their jaws, was a row of human skulls. *Herr* Schmidt said, "Filipovic—are you a Yugoslav?"

I said, "Yes."

"Serbian or Croatian?"

"Croatian."

He stood up, put his hand on my head—my hair was cut so short I was almost as bald as he was—and turned it so my profile was presented to the four officers. He said, "Gentlemen, here you see an example of the perfect Slavic skull."

When he twisted my head around I was facing the row of skulls and while he went on to describe the differences between Slavic and Aryan skulls, I stared at the ones on the shelf, thinking, "This Hun is going to boil my head so he can put me in his collection."

Now *Herr* Schmidt was standing in front of us in the classroom. He positioned himself on the podium and addressed us formally:

"Young men, you are very fortunate to have been selected for this special training program provided by the *Reich* for only a very small number. But the failure to respond with the *Führer's Gruß* is a very serious offense. I will give you one last chance to give me the proper greeting." He raised his arm in the Nazi salute and said, *"Heil Hitler!"*

The room was silent. Jan had gotten up and was standing in the first row, his head bowed and bloody. He didn't move, no one moved. If just one of us had returned the salute then we'd all follow and be saved. But nothing happened, we just stood there.

Herr Schmidt's face, which was always red, got even redder. He dropped his hand and walked out, banging his heels on the floor. The teacher went to his desk, gathered up his papers and said in an arrogant voice, "There is nothing more I can do for you." He walked out. After a few minutes we all sat down and started talking about what had happened. We wondered why they had tried to make us give the *Führer's* greeting when they already knew what we thought of them. It almost seemed as if they wanted us to defy them.

Two guards came in and shouted, "Stand up! First row, march out! Second row, march out! Third row...." They marched us back to the camp but instead of going to the barracks we'd been sleeping in for the past three weeks, we kept marching to another one further away. They herded us in and left us there. There were no bunks, just a concrete floor covered with straw. This didn't look very nice. There was no food at noon and a couple hours later the guards came back and shouted, *"Alle raus, raus!"* We lined up outside. "Take off your clothes, strip to the skin, everything!" The guard showed us how to fold our clothes in a bundle. We marched to another building, naked, carrying our bundle of clothes. It was cold, freezing, but I was too scared to feel it. We turned in our bundles and were given different clothes, a blue shirt, pants and jacket, all with patches and rips sewn up. I knew these clothes. It was the uniform the people wore who worked down below. We each got a blanket and were marched back to the barracks. It got dark. No supper. There was nothing to do but make a bed with the straw on the floor and the blanket. Some of the guys talked about what was going to happen to us. They were all from Holland and Belgium, all at least seventeen or eighteen, two or three years older than me. They reminisced about their homes, their families and friends, the places they'd gone dancing and met girls, but after a while there wasn't much talk. We were all very subdued. I lay in the straw by myself. I knew the next day I was going to be put back to work again, hard work, and I knew I could stand that. But I also knew this time I wasn't going to get a job on top of the shaft. This time I was going down—we all were—we were going down into the coal mine.

"Alle aufstehen!" the loudspeaker shouted.

I jumped up, ran out to the latrine, did my business, washed my face—cold water, nothing but cold—and ran back. Everyone ran, you ran to keep warm. Breakfast was a hunk of black bread that felt like sawdust in my mouth and a scoop of ersatz coffee made from tulip bulbs. We marched out of the camp. They tried to keep us in formation but as soon as we got out of the gate and on the main road with the people from the other barracks, everyone clattered their wooden shoes on the pavement like a herd of cattle. We walked, hundreds of us streaming toward the coal mine complex, through the countryside, past farmhouses, through the outskirts of town, past houses and shops, along the trolley line where the ordinary Germans stood waiting to go to work and school. We went right into the center of Kohlscheid and through the huge iron gate into the complex.

We were divided into groups according to where we worked. I was told to go to a one-story building to get my acid lamp. Inside, I waited in line, the smell of acid burning my nose. I told the man behind the counter my number and he went down a long rack, yanked out a lamp and brought it back. This guy had a bad job, as I learned later when I had to do it myself. He had acid burns on his hands and holes in his clothes from the spilled acid. The worst part was grabbing the lamps from the racks where they were recharged because every time he touched a lamp he got a shock.

I hung the lamp around my neck like everyone else and we climbed up six flights of stairs to the top of the shaft to wait for the elevator. All around us was the noise and racket of the operation. I could see the place where the lorries came up. Some other guy was standing there catching them now. When the elevator came up we piled on, twenty guys with the foreman, they lowered it a couple meters to the next cage above, twenty more guys piled on over our heads, it jerked down again and the third cage was filled. The bell rang twice—two rings was the signal to go slower because the cage was carrying people instead of material—but nevertheless, we shot down into the dark. The cage rattled and shook with the wind rushing past, there was a flash of light when we passed the first level, then dark again, darker than night, absolute black. We kept going down, down to the lowest level, the seventh, a thousand meters below ground. When I stepped out I felt like I was in the center of a small village, a dark village with a low ceiling. There was activity everywhere and tracks going off in all directions. A small locomotive hauling empty lorries drove up and I was told to get in. As we rode away I could see what looked like "shops" along the main street of the village—the carpenter, the machinist, the blacksmith, the stables for the horses they used to pull heavy loads. Then it was dark again. All I could see in the dim light on the front of the locomotive were the walls of the tunnel. Now and then the train stopped to let a couple guys out. I rode for a good half hour before we stopped and the driver shouted, *"Rivier vier!"* Section four. I got out with the miner I'd been assigned to. "Turn on your light," he said. "Follow me." We walked through a tunnel just high enough to stand up in. "Has anyone trained you?"

I said no.

"Then listen to what I say." He was a young German, very gruff and unfriendly. "You're my *Rutschenhelfer.* You're going to work on my *Schüttelrutsch."*

The *Schüttel*. I'd heard of it. You couldn't help hearing about it because it was such a lousy job. We got to his shaft. He hung up his jacket on the end of a shoring timber and I did the same. In the light from our lamps I could just see the end of the *Schüttel*, a flat metal trough with sides, jutting out the end of the shaft. There was an empty lorry parked by the end. The miner turned on a switch and the *Schüttel* started jerking back and forth, making a racket.

"The coal comes down here and when the lorry's full they push it out to the main tunnel," he shouted. "Come with me." He bent down and disappeared into the shaft. This one was too low to walk in so I crawled on my hands and knees alongside the *Schüttel* for a hundred meters or so till we got to end of the shaft where he was digging.

"When I chop out the coal and the stone, whatever I dig, you shovel it on the *Schüttel*," he shouted. He had to shout because of the noise of the *Schüttel*.

"That's your job. Don't get behind!"

He gave me a short-handled shovel and pads for my elbows, grabbed his air hammer and started digging.

The noise in the shaft was incredible. An air hammer cutting out coal and stone, the sound of it hitting the *Schüttel* when I shoveled it in, the racket of all the coal and stone bouncing a hundred meters down to the lorry while the *Schüttel* jerked back and forth, and the constant roar of the engine running the whole contraption. The dust was so thick it was sometimes hard to see or even breathe. In order to shovel I had to lie on my stomach and the farther ahead of me the miner dug the harder it was because I had to throw the coal backwards and up on the *Schüttel* at the same time. I also had other duties. When the *Schüttel* was too short to reach where he was digging, I had to crawl back out of the shaft and get another section, a metal trough three or four meters long, then drag it over the top of the *Schüttel* to where he was digging and hook it on. I also had to put up the shoring timbers. They were cut up aboveground and the ends were notched so the horizontal plank fit into the vertical one and I could slide them in place. I had to shore up behind the miner as he dug out the coal so if the roof of the shaft caved in it would settle on the timbers and not fall on him, or me. All the time I was doing these things the miner was chopping like hell because unlike me he was paid and the more coal he chopped the more money he made.

At noon the miner stopped work and crawled out to eat his lunch. I crawled out too. The other miners were sitting around by the end of the shafts eating the sandwiches and coffee they'd brought from home. I noticed a couple other prisoners, guys like me, eating a piece of bread they'd saved from breakfast, but I hadn't saved any. I sat down by the wall while the others ate and thought about my hunger. All I could think about was food but I got nothing to eat till we marched back to camp at the end of the day. Supper was another hunk of sawdust bread and a bowl of soup, mostly water with a few pieces of cabbage and potato floating in it. The next morning I was so hungry there was no way I could help myself from eating the whole hunk of bread, so once again I had nothing for lunch. Next morning, up at six, march to the mine, work ten hours, march back, sleep. The days were the same, seven days a week, and my preoccupation was with one thing, food, always food. I was constantly hungry, all day, all night, even after I ate. We all were hungry, it was the one thing on everyone's mind, food. Food and cold.

Two weeks after I started work on the *Schüttel*, the *Steiger* came around during lunch break—as usual I hadn't saved any bread and had to watch while the coal miners ate their sandwiches—and assigned me to a different job. I was to push the lorries full of coal from the end of the *Schüttels* up to the turntable and then out to the main line where the locomotive hooked them up to go to the elevator. The Dutch guy who'd been doing this job was too weak to push them and the miners were complaining they were losing money because they had to wait. So the *Steiger* put me in charge of six *Schüttels*. I was glad to be out of the shaft, but right away I saw why the Dutch guy couldn't do it. The wheels on the lorries were so bad some of them wouldn't have rolled down a hill, and so much coal and stone had fallen on the tracks the wheels were always getting jammed. So even if I could push the lorry, the wheels would get stuck. Then I had to get a shoring timber, wedge it behind the lorry and sort of nudge it along the track. The turntable got stuck too from the coal that fell off the lorries and I had to use a shoring log like a crowbar to pry it around. All the time I was trying to move the lorries and crank the turntable, more coal was rolling down the *Schüttels*, filling more lorries. It wasn't as noisy and dusty as shoveling up in the shaft, but it was harder, heavier work, and I could barely keep up.

Less than a week after I started the new job, I sat up in the straw one morning and felt faint, as if I was going to pass out. I wasn't sick, it was my hunger. I was losing weight, losing strength. I couldn't do the job on the food I was getting. After I ate my hunk of bread and drank my coffee I felt better. Still hungry, but not as weak, and on the way to the complex I ran over with some of the others to see if there was anything to eat in the trash cans the Germans had left out to be picked up. Not everyone dove for the trash. Some were too proud, they had just enough pride left not to eat garbage. Others were afraid. If they got knocked down digging in the trash cans and the guard came over and beat them, they might not get up. My head was down in a can and I was grabbing for something when I got hit in the back. I jerked myself up but it wasn't a guard, it was some German kids from the town throwing rocks. They were shouting and cursing us, calling us dirty swine. I hated them even more than the guards. We were so hungry we would eat garbage, so to them we were pigs.

Later that morning down in the mine it happened again. A lorry got stuck and when I bent down to pick up a shoring log I felt dizzy, like I might faint. If I didn't get something to eat, I couldn't do the job. I knew what that meant. Every morning when we stood formation outside the barracks, the guards went in and shouted at the ones who were too weak or sick to work, "*Faulenzer!* Get up!" and if they didn't get up, the guards would shout, "You better not be sick Friday!" On Friday the truck came to pick up the ones who had died in the barracks during the week and if you couldn't get up they dragged you out too, dead or alive, threw you on the truck and drove you down to the boxcars on the siding in the complex. For the first time since I'd been caught, I was afraid I wasn't going to make it. I had to find food. But there was no food.

I was standing there, leaning against the lorry, when I saw something. Hanging from one of the shoring timbers was a coal miner's jacket, and sticking out of the pocket I saw the corner of a newspaper. As soon as I saw that I knew what it was. I walked up to the jacket, stuck my hand in the pocket, pulled out the package and unwrapped the newspaper. There was a sandwich, two small ones, on *Kommissbrot*, army bread, with sliced egg. I ate them both. The coal was pounding down the *Schüttels* and filling the lorries I was supposed to be watching but I didn't care. Let 'em go to the balls, I cursed them in Dutch. I was thinking about something else now. I went down the tunnel and found another jacket hanging from a shoring log. I slapped the pocket—WHAP WHAP—to feel what was inside. This miner only had one sandwich but it was on *Bauernbrot,* farmer's bread, with a piece of meat. I ate it.

By the time I got back to the shafts, the coal was all over the place, enough lying on the ground around the lorry to fill up another one. The miner was going to be pissed off when he crawled out for lunch and saw all the coal he'd dug out lying on the ground at the end of the *Schüttel*. I also knew there were going to be two other miners who were pissed off when they reached in their jackets for lunch. But I didn't care. I had something in my stomach and I had my courage back. I started pushing the lorry but it wouldn't budge. It was trapped by all the coal on the tracks and more was coming off the end of the *Schüttel*. I was going to have to clean up the mess before I could move the lorry, so I crawled up in the shaft and rapped the *Rutschenhelfer* on the back. He was some Polish guy and I told him, "Shut off your *Schüttel!* There's too much coal coming off!"

The guy looked at me like I was crazy. "You want to shut off the *Schüttel,* tell your *Vorarbeiter*. Don't tell me!"

The *Vorarbeiter* was the foreman.

"Just tell the German to quit chopping while I clean it up!" I shouted at him.

I crawled back out and went to the next shaft where I did the same thing only this time I flicked the switch and shut off the *Schüttel* myself. I didn't wait for an answer either. I just told them to stop and I could hear the miner way up the shaft cursing. "What the hell's going on?" By the time I had crawled in and out of the other four shafts some of the miners and helpers were crawling out too. It wasn't noon yet but if they couldn't work they were going to eat lunch. The miners didn't look too happy seeing all that coal all over the place but they knew they'd been hitting soft coal and really breaking it out, so they just muttered a few curses and ignored me. I made myself busy clearing the coal from the tracks and at the same time I was aware of the two miners whose sandwiches I'd eaten. Other people walked back and forth in the area so it could have been anyone stealing their sandwich. The guy with the egg sandwiches on *Kommissbrot* went to his jacket, reached in, found nothing, turned it around and tried the other side. In that pocket was his *Puer*, a flask the miners carried coffee in. He tried the other pocket again, then went over and sat down by the wall. He didn't say anything, he just drank his coffee and sat there looking thoughtful. I kept working.

A few minutes later I heard someone shouting up the tunnel. "Some filthy swine—*Diese dreckigen Ausländer haben meine Mahlzeit gestohlen!*—The foreign filth have stolen my lunch!" He came stomping up to where the others were eating and the one guy is drinking his coffee, and he's screaming, "These filthy foreigners *haben meine Mahlzeit gestohlen!*" Meanwhile, I'm making myself busy with one of the lorries, picking the coal out of the wheels, and the first guy, the one with the egg sandwich, said, "You think so? My lunch is not in my pocket either! *Aber ich dachte, ich hätte meine zu Hause vergessen*—I thought I left it at home." The other guy, the one who had the good sandwich, the one with meat in it, said, "I'm sure I brought mine this morning. Ach, these filthy foreigners, they must have taken it." He went back up the shaft, pounding his feet.

I had pushed the lorry onto the turntable when the *Steiger* walked in from the main tunnel. He saw everyone sitting around and said, "What's going on? Why aren't you working?"

"The lorries are full," one of the miners said. "The guy can't keep up with us."

"What the hell is wrong with you?" the *Steiger* said to me. "What are you, a *Faulenzer?*"

"No, I'm not lazy," I answered him. "There's too much coal coming out. I can't keep up with it."

He started cursing me but a couple of the miners told him they'd been hitting soft coal and it was too much for one guy to handle—not because they had any love for me, they just wanted to get paid for all the coal they chopped out.

"Okay," the *Steiger* said to me. "You get back to work. I'll get some help for you tomorrow."

The next morning when I got to *Rivier Fier* there were two young Dutch guys assigned to help me move the lorries. I showed them the job and gave them two *Schüttels* apiece to watch. Then I said, "If I'm not here you have to watch mine too." They were scared, they didn't ask where I was going. "Just make sure you take care of my lorries, push them out when they get full or there'll be trouble."

That day began what became my routine. I got to work about 9:00 or 9:15, depending on how long it took to wait for the elevator, and about ten o'clock when all the miners were up in their shafts chopping like hell and the *Schüttels* were going, I left the lorries and went up one of the tunnels for a sandwich. The miners all hung their jackets on a shoring log near the shaft before they crawled in, so I felt along the wall for the fabric. As soon as I got near a shaft and hit cloth, I stuck my hand in the pocket, pulled out the sandwich and ate it. After a week or so the flow of coal slowed down and the Dutch guys were reassigned somewhere else, but that didn't stop me. I kept right on stealing sandwiches every day. If there was too much coal coming out, I just shut off the *Schüttel* and shouted up the shaft, "I've got too much coal, stop for a minute!" The miners cursed at me but they stayed up in the shaft because it was such a pain in the ass to crawl all the way out. Besides, I wasn't stealing their sandwiches. After the first incident I'd learned to be careful—not to shit in my own backyard. As time went on I also increased my territory because the problem of stealing sandwiches became well known in that part of the mine. The Germans knew very well it was one of the fucking foreigners. But I knew my way around, and I became very cocky. When I left the *Schüttels* and went up the tunnel, I took off my lamp. It was dark down there—it wasn't Broadway—and as I approached a shaft I ran my hand along the rock till I came to something. If it was cloth, I slapped the pocket—WHAP WHAP—I frisked it, till I hit something. If I felt the *Puer* I turned the jacket around. I didn't want coffee, I wanted the sandwich, so I'd hit it—WHAP WHAP—to see if something was in there before I committed myself to reaching inside.

It wasn't long before the Huns got so mad they started laying for me. But what they did was so stupid they didn't have a chance.

They'd stop working and hide near the end of the shaft, waiting for me to come along and grab their jacket. But if I came to a shaft where the miner wasn't chopping, I just kept going. It was easy, if the *Schüttel* was going and coal was coming off the end, the miner was up there digging. If it was off or no coal was coming down the chute, he was watching. Plus, I never went to the same place two days in a row. I spread it around. I was even kind, I'd come to a jacket and say to myself, "I took his lunch last time, I'll let him eat today." But I got ornery too. Sometimes after I took a sandwich I'd turn the jacket around, pull out the *Puer* and pour the coffee on the ground. "Let 'em go to the balls," I'd curse them. "Let him wait till he gets home tonight to get something." As time went on I kept increasing my territory so I was going further and further for sandwiches. I might go three, four hundred meters, as far as a kilometer from my shaft. Some days I'd even steal more than I could eat. I'd take a bite and throw the rest down a dead shaft, just to make the *Mof* go hungry. That's what having a full stomach does to you. When you're hungry, you're subservient to them. You get something to eat and your attitude changes. You say fuck 'em. Let the rats eat their lunch. That's what happened, I became ornery. Because now I knew I could survive. I'd found something to eat and no one was going to put me in the boxcar.

CHAPTER 2
KOHLSCHEID, GERMANY; WINTER-SPRING 1944

For the next year and a half I worked down below. Occasionally I was pulled off to do another job, like unloading logs, or tending the lamps in the acid lamp shed, but most of the time I worked down in the mine. At first, I didn't think about escaping from the camp because even if I could get out where the hell was I going to go? I would still be in Germany and if I could get back to Holland there were not only the German soldiers but also the *NSB*ers, the Dutch collaborators. A punk kid like me with no papers couldn't just walk down the street. So I was just getting through every day, waiting for something to happen, and all I thought about was food—food and how to stay warm. That was your life.

When it got cold in the winter, more people were carried out of the barracks, and I looked around for a warmer spot to sleep in. I was sleeping in the coldest place, out in the center where I also got kicked in the head by the people sleeping in the next row. We slept on the floor in some straw with one blanket, rolled up next to each other. One freezing night I woke up and the Dutch guy I slept beside was cold. I shook him. He was gone. I rolled over against the guy on my other side. In the morning when I woke up he was dead too. After we ate I moved my blanket and bowl over beside an older Polish man named Bogdan who had a place by the wall. It had been a bad night, very cold, and the guy beside him hadn't made it either.

I met Bogdan almost as soon as I came into the barracks because he always asked any new people if they had heard of his family, especially if they were from Poland. I wasn't, but I knew some Polish I'd picked up from my friend Vwadzu at home, and I also spoke English with Bogdan. I'd learned English in school and he'd been a teacher in Poland before the war; he was educated and spoke several languages. So sometimes at the end of the day in the barracks we talked, and his family was on his mind all the time. The last time he'd seen them was when he got picked up on the street in Poland more than a year ago and his hope that they might be alive was what kept him going. Some people worried all the time about what was going to happen—what if they couldn't work, what if they fell and couldn't get up, what would the guards do to them? I hated to hear people talk like that because it made them weak. I never thought about what might happen; I just did what I did every day. With Bogdan it was his family, he talked about his wife, his children, his parents, and he talked about his hunger. When we rolled up together to sleep at night, he complained about it. I didn't say anything. I was afraid if I said anything to anyone about stealing sandwiches word might get out and that would be the end of my operation. Because if I hadn't known it before, now I knew—when you're hungry you're not an honorable person. But I could see how desperate Bogdan was getting. All he talked about was his hunger.

One day down in the mine after I grabbed a sandwich, instead of eating it I shoved it under my shirt. That night after they shut off the lights in the barracks and we rolled up to sleep, I reached in my shirt and poked Bodgan. "What is it?" he said. I put the sandwich in his hand. At the other end of the barracks someone was screaming and cursing. It happened sometimes, people got on each other's nerves, there were fights. When it got quiet again I heard Bodgan whisper, "Josko, thanks. God bless you." After that, when his hunger got bad and he complained he couldn't work, I'd say, "Wait till tomorrow," and bring him a sandwich. He never asked how I got it, he just took it and ate it.

The routine was the same, the same fucking thing every day—get up, eat, stand formation, march to the coal mine, work, march back, stand formation, eat, sleep. Sleep was the only thing we got enough of, and you just passed out, so fucking tired. The only thing that changed was what was in your mind. Other people might have thought, like Bogdan, about their families, the people they were missing. I thought about how to escape. I thought about it every day but it was just a release because, even if I got out, as long as the Germans controlled everything there was no place to go. But in the spring of 1944, when I'd been in the camp about a year and a half, I overheard some of the German miners talking about an invasion. It was during their lunch break and one of them had heard a rumor the *"Amis"*—the Americans—were going to try to break through the *Westwall*. The *Westwall* was the German defense line, the original Siegfried Line that Hitler rebuilt. It ran all the way from northern Holland down along the border of Belgium and France to Switzerland, and for years the *Moffen* had bragged that nothing could get through it. Perhaps they really believed it too, but I could see the miners were worried, and I thought if the *Amis* could do it, if they could break through, the end would have to come soon.

Then one day it seemed as if all the *Moffen* were talking about it. "Invasion, invasion, invasion," that's all you heard. Every day for weeks they talked about when it was going to happen. "Next Tuesday." "Saturday." "Another week." Then when nothing happened they started making jokes, laughing about it. "There isn't going to be an invasion." But finally, in June, the invasion began—what in England and America was called D-Day. The German miners just called it the invasion, *die Invasion,* and even if I hadn't heard them, we all knew something was going on because the bombing had become just ferocious. Ever since I'd been in the camp the American and British bombers had been flying over and hitting places in Germany—*Amis* during the day, British at night—but now they were up there in almost a steady stream, a constant droning and roaring, three or four huge air raids a day and then several at night. They hadn't bombed Kohlscheid yet but they were dropping them closer and closer all the time. At first, when the invasion began, the Germans said, "We'll push them back into the sea," but after a couple weeks they quit talking about how the German army was going to defeat the enemy, and then I knew the British and the Americans had made it. The "enemy" had broken through.

I decided I was going to get out. If I could escape now, I figured I could hide out in Germany in the countryside, in the fields, the tall grass, wherever, and wait till the *Amis* came. It was warm, things were growing, I could survive. It wasn't just that I wanted to get out, I was also worried about what might happen if I was still a prisoner when the Huns got overrun. The guards had started hinting what they were going to do to us. "If you think the *Amis* are going to win the war, don't be so elated because if they get this far you'll never see them." The guards were just old men, too old for the army, and maybe they threatened us out of frustration that their army was being pushed back and all the bombs were dropping, but nevertheless I knew they might do it. Because to them we were nothing. We were shit, the lesser people, *die Untermenschen.* But as determined as I was to get out, at the same time I knew I couldn't hide that long in Germany waiting for the *Amis*. I had to be patient. So all summer I listened to rumors, to what the coal miners were saying, to learn how close the *Amis* were getting to us. From what they said I knew the fighting was rough, very rough, and by late August I was getting anxious. When the hell would the *Amis* get here? By now they were in Belgium—closer but still a long ways off—and I could see the *Moffen* were tense. Their anxiety was building every day. The bombers were up in the sky all the time, every day, every night, and I was worried they might drop something on us. I couldn't wait any longer, it was time to get out.

By chance there was someone in the barracks I slept in who was also from Frankstraat. Leo Timmerman was a Dutch guy a couple years older than me and not only was he from my street but his younger brother Scheng had been one of my best friends. Leo had been picked up in Heerlen one day when the *Moffen* were out looking for young guys, just grabbing them off the street. He'd been in the camp almost as long as I had and although he was in pretty bad shape, he had survived. One night after we ate the bowl of "soup" I went over to his place in the corner of the barracks and said, "Leo, I'm going to get out of here. You want to come?"

He looked at me like I was crazy.

"Get out? How're you going to do it?"

"I don't know yet," I said. "I'll let you know."

I had an idea but I didn't want to tell him, not till I was ready to go.

"Okay," Leo said. "If you think we can do it."

"Yeah," I said. "We can do it."

At home I was always the leader, like when we fought the rich Dutch kids from the good neighborhood. So Leo didn't know how I was going to get out but he believed me. A few days later, he came up to me when we were marching back to the camp and said, "Josko, are we going to get out? You still going to try?" He was getting anxious.

"Soon," I told him.

It was early September by now and I was just waiting for the right night. It came a couple days later. When we came up from the coal mine I could tell right away there had been very heavy air raids that day. To the south the city of Aachen was burning and in the other direction there was smoke over Herzogenrath, a town on the border with Holland. It was a nice, warm evening with no sign of rain and when I saw how badly everything close to us had been hit, I thought—this is the night to do it.

At ten o'clock the lights were shut off in all the barracks at once. There was a loud CLANK and everything went dark. Outside, through the narrow windows that ran along the base of the wall, I could see the spotlights in the guard towers around the camp—they were still on. That was the routine every night. I lay in the straw waiting. About an hour later, I heard what I was waiting for. The sirens started up signaling another air raid and almost immediately the Huns shut off the spotlights. I got up and walked in the dark through the barracks, past all the bodies sleeping on the floor, and found Leo. I grabbed his shoulder and shook him, woke him up. "Leo, get up."

"Josko?"

It was so dark he couldn't see me.

"Come on," I said in Dutch. "I'm leaving."

"Tonight? How come tonight?"

All of a sudden he's reluctant.

"Come on," I said. "Let's go."

"Are you sure? Maybe we should wait."

"Yes, I'm sure. Come on."

Leo got up. Scared as he was, he didn't want to get left behind.

I walked back down to the other end of the barracks to the door and as I passed the place I slept I heard Bogdan say, "Josko? Where're you going?"

I didn't say anything.

The door to the barracks was open. They didn't bother to lock it or put a guard outside—where the hell was anyone going? I opened it and looked out. The yard was dark and empty, all the lights were shut off.

"Wait," I told Leo. "Wait till the next wave."

I didn't have to tell him to wait. He wasn't going anywhere until I did.

A wave of bombers had just passed over and we were going to have to wait a few minutes for another. But it wouldn't be long. I could already hear the next wave coming in. It sounded like thunder, in the distance.

"Where're we going?" Leo said.

"Over the fence," I said.

The first fence was about a hundred meters away, across the yard. It was made of strands of barbed wire strung on concrete posts three or four meters high, about twice as high as my head.

"We'll get fried, Josko—it's electric!"

"I think they shut off the juice," I said. There were signs on both sides of the fence warning it was electrified but I figured when they turned off the spotlights during the air raid they shut off the power to the fence too. I didn't know that for sure, but it was the only way out. We had to take a chance.

The next wave was rolling in. I could see the shapes, the silhouettes of the bombers, over the roof of the barracks and towers. The stars were out, a clear night, and a moment later there was no doubt where the planes were. Right overhead. The roar was tremendous. Everything shook, the barracks, the ground, it went up through my feet to my head.

"Come on!" I shouted at Leo.

I ran across the yard and when I got to the fence I didn't stop, I just reached for the wire and started climbing. If the juice was on I never would have known what hit me but I was still moving, climbing as fast as I could, trying to grab the wire between the barbs. I got stabbed a few times but I was in too much of a hurry to be careful. Going up I was okay, but at the top, several strands of wire were strung on struts angled up and in, toward the camp. I reached up, grabbed the highest strand, and pulled as hard as I could. At the same time I swung my leg up in the air, over the wire. The barbs tore my pants and dug into my skin but my leg was over the top. I hauled myself over the angled wire till I was on top of it, sideways. The barbs were stabbing me everywhere and now there was a lull. The wave of bombers was passing. It seemed like everything stopped. It was quiet. I kept working myself over the angled wire but now I could hear my own noise, the wooden shoes scraping the wire, the barbs tearing my clothes. I could hear Leo too, he'd made it up to the top, but we were taking too long, making too much noise. If another wave didn't come in pretty soon they would sound the all-clear and turn on the lights and the power. I felt for the wire on the other side with my foot and tried to turn myself around to climb down. Now the barbs were holding me back, snagged on my clothes and skin. I was stuck, hung up in the air. My foot slid off the wire. Then I caught a barb with the shoe and swung my other leg clear. I was over, climbing down the other side and I didn't care where I put my hands, I just grabbed and dropped as fast as I could. Leo came down almost beside me. We ran to the next fence. Another wave of bombers was coming in.

The second fence was barbed wire too but instead of strands it was in coils, shoulder-high, unrolled on the ground. It wasn't meant to stop you but to slow you down, get you tangled up in the coils. From looking at it when I walked to the coal mine I figured if I was patient I could maneuver my way through, but it was hard to be patient with the roar overhead. I felt in the dark for the wire, grabbed a strand and pushed it to the side while I stepped through. The barbs snagged my pants, I yanked my legs through, all the time thinking, "Move your ass! Move your ass!" It was taking forever to work my way through the coils. I realized I was moving sideways. Then I reached into the dark and felt nothing. I pulled my leg free and ran up the road. I was out.

"Leo!" I shouted over the roar.

"Josko!"

He was behind me.

"Come on!"

We ran across the road into a field, and kept running through the field till we came to a path. Then we followed the path. It was a good clear night, the stars were bright and I could see the path ahead of me, and the shapes of trees. We ran for fifteen, twenty minutes, a good kilometer or two, just running to get away from the camp, and when I saw a house up ahead on the path I slowed down. Leo caught up with me.

"Where're we going?"

"I don't know," I said.

We were out, that's all I knew. But Leo was beat, he had to sit down. I was tired too but Leo hadn't been stealing sandwiches and eating like me. He was in bad shape. While he sat there resting I looked around, trying to get oriented and figure out which way to go. We had come out of the camp on the opposite side from the road we took into Kohlscheid every day. Because of the air raid all the lights in the the town were blacked out but I knew where it was and I knew which way we were running, east. When we got out I didn't think about which way to go, I just ran like hell to get away from the camp. Now we were going to have to get off the path and cut through the field because we couldn't run right past the house. Then I had an idea.

"Leo—let's break into that house."

He looked at me like I was crazy.

"We've got to get rid of these clothes," I said. "Maybe we can find something to wear."

Our clothes were ripped and torn from climbing the fence and besides, these were not clothes we could be seen in once it got light. Everyone knew the blue uniform. They would know right away we'd escaped from a camp.

"Suppose there's somebody in there," Leo said.

"If anyone's there, they're in the cellar."

The house was dark, blacked out like everything else because of the air raids.

"Even if they're in the cellar, they'll still be able to hear us," Leo said.

"We'll wait till the next wave comes in," I said.

Another wave of bombers was on its way, still a couple minutes off, but I'd guessed right, it was a huge bombing raid tonight. We rested till the planes were right overhead and the ground was shaking. The noise was incredible, from the bombers above and the German anti-aircraft guns blasting all over the sky.

"Come on!" I shouted to Leo.

We ran to the house, around to the front door, and I grabbed the handle. It opened right away. Someone was home. I stepped inside, into a foyer, and said, "Hello?" No one answered but the roar of planes was so loud by now even Leo couldn't hear me. I walked along the foyer feeling the wall for a light switch. My hand came to a porcelain knob and I turned it. There was a clank, a vibration, and the lights came on. I was standing in a combination living room-dining room. To my right was a door slightly ajar. I found another light switch just inside the door, went in, and there was a big wardrobe right there in front of me. It was full of clothes, men's and women's both. I grabbed a sport coat, a nice brown and gray checked jacket, and a pair of pants. In the bottom of the wardrobe were shoes and I grabbed a pair, the best-looking ones. There was an overcoat too but I didn't bother with that.

"Grab some clothes," I said to Leo. He was standing there watching me.

Beside the wardrobe was a dresser and inside were shirts, underwear, socks, everything. I started tearing my blue uniform off.

The action outside was still very heavy. The roar from the planes hadn't let up and the anti-aircraft guns were still blasting like hell, so when I saw a bathroom off the room with the wardrobe I decided to wash up too. Our hands and arms and legs were all cut up, scratched and bloody from the barbed wire, so Leo and I both undressed and cleaned up while the bombers and guns pounded outside. My new clothes fit like they'd been made for me, even the pants, and the shoes were nicely polished and they fit perfectly too. I had on the guy's best outfit and was a hell of a lot better dressed than I'd ever been in my life. Leo had found another jacket and pair of pants, a little short for him but not so small that they looked like they weren't his, and he looked pretty sharp too. We rolled up our camp clothes and left them with our wooden shoes in the shower. When I closed the front door behind us, another wave of bombers was coming in.

We started running.

We were still running east, away from the camp, and when the path ended at a road we turned right, to the south, still heading away from Kohlscheid. The road was paved and I could see a few blacked out houses scattered around but at least we were getting away from the outskirts of Kohlscheid. It was starting to get darker, some clouds had come in, but the planes were still up there and off to the north and west the sky over Herzogenrath was just raining with bombs. The Huns' searchlights were crisscrossing the sky and occasionally I saw the outline of a plane. The anti-aircraft shells were so loud I couldn't even hear the sound of our shoes on the road, so it took me a minute to realize I was running alone. I looked back and saw Leo had stopped. He was whipped, out of breath, too tired to run any more. We were by the side of a wheat field and I couldn't see any houses so I figured it was okay to rest for awhile. We sat down by the side of the road. The ground was vibrating from all the action but we were safe for the moment. They weren't bombing Kohlscheid but they were pounding the hell out of Herzogenrath and no one was going to be out during a raid like this. We sat there watching the bombers. The lead plane would drop an illumination flare that lit up the sky like a huge Christmas tree, a signal to the planes behind it to drop their bombs. It was a beautiful sight but as soon as you saw it you knew to watch out, death was behind it.

"Where're we going, Josko?" Leo said.

I didn't know. My idea had been just to get the hell out of the camp and then find someplace to hide, but sitting on the edge of the wheat field with all the bombs and shells blasting I realized it wasn't going to be easy. Everything around was either farms with houses or villages with more houses. It was all settled and cultivated and we didn't know the neighborhood, the places to hide. I was sitting there thinking when I noticed the jacket I was wearing. It gave me an idea.

"We're going to Holland," I said to Leo.

"Holland? You mean home?"

"Yeah."

"How?"

Leo didn't believe it. I wasn't sure either, but now that we were out of the camp it was obvious to me that we couldn't stay here in Germany. Even if we found a place to hide there was so much bombing no place was safe. Before, I hadn't even thought about crossing the border because we were wearing the blue prisoners' uniform but now if anyone saw us they wouldn't know right away we'd escaped from a camp. Our hair was usually cut almost bald but we hadn't been clipped for a couple weeks and there were plenty of Germans running around with short hair too.

"I think if we turn right again as soon as we come to a road and keep going as far as we can, we'll come out south of Kerkrade." Kerkrade was a Dutch town just over the border and although it was a distance from Kohlscheid, it wasn't that far from Heerlen.

"The border's all fenced, they've got guards," Leo said. "We can't get through at Kerkrade."

"Maybe if we stay south of Kerkrade—I don't think it's fenced further south."

It was all farmland around Kerkrade and would be safer, but it meant we might have to circle around Herzogenrath. I wasn't sure. I knew where Herzogenrath was from the bombs and smoke but wasn't sure if it was north or south of Kerkrade. I didn't want to get lost. But Leo didn't object to the idea—he didn't say anything, and when I looked at him again he was lying in the wheat, asleep. I was tired too so I sat there resting for half an hour or so till I realized no more planes were coming in. The raid was over and now the bombers were flying back the other way. I heard a siren, quite a ways off, sound the all-clear.

"Wake up!" I gave Leo a punch. "If those people in the house come up from the cellar they're going to call the police. Someone might be out looking for us."

We started walking again and when we came to an intersection we turned right, toward Holland. It was a dirt road but a good one, well-traveled, and I thought maybe if we kept moving we could get to the border before light. That was just a hope. I knew it was too far. But all we could do was keep walking, and tired as Leo was, he kept going, kept moving. When I finally saw the first light in the sky behind us, we were still on the same road. The sky had cleared, it was going to be a nice, sunny day. I could see a few houses in the distance on either side of the road but couldn't tell how far we'd come or where we were now. I tried to orient myself by Herozgenrath which was in a long, shallow valley so everything in that direction sloped downhill. Smoke was still rising from the town. I was right, it had been hit very hard during the night. Finally, up ahead of us, we saw a paved road crossing the one we were on. The dirt road continued past the intersection but ended at a farmhouse, so we were going to have to turn one way or the other.

"Let's take a left," I said to Leo. "As soon as we come to another road we'll go right, and if we don't come to a road we'll cut across the fields."

The paved road was a main highway with streetcar tracks running down it and I figured this must be the road to Aachen, the major city in the area. Before the war Mama had sometimes taken the trolley from Heerlen across the border to Aachen to go shopping, and I had come with her. But the paved road was a bad place to be, it was too public, and I was already thinking we should cut into the field when I heard something coming behind us. It was an open car with no top, a German army Volkswagen, like a Jeep, with several men in uniforms riding in it.

"What is it, Josko?" Leo said. "You think they're police?"

"I don't think so. I think they're German army."

Whoever they were, it was too late to jump into the field. They'd seen us too. The Volkswagen came closer without slowing down, we kept walking without looking at them again, and when it came to us it zipped right by—four Huns riding along in black uniforms outlined with silver. The *SS*. They kept going. We kept walking.

The tires screeched. The Volkswagen stopped. I could hear the gears grinding and it backed up.

"If they ask us anything, let me do the talking," I said to Leo. He had already made up his mind to shut up and let me deal with them. I would lie and he would swear to it.

They stopped beside us and the one beside the driver got out. He was young, but an officer, missing an arm. One sleeve of his uniform was tucked in a pocket.

"Guten Morgen," he greeted us briskly. "Where are you going?"

"We're going to Aachen," I said in my best polite German. My accent sounded like a native's by now.

"Aachen?" He raised his eyebrows when he heard that. "Aachen is far."

"Yes, I know," I said.

"Show me your papers," he said.

"We don't have any—they got burned." It was the first thing I thought of.

"Burned?"

"Yes. We live in Herzogenrath and got bombed out last night. We lost everything except our clothes so we started walking to Aachen because I don't think there'll be any streetcars today."

"No, I don't think so," he said. "Why are you going to Aachen?"

The officer was staring at me and I stared back at him, right in his face. Everything he said was very fast, very brisk, that was the Nazi manner.

"My aunt lives there," I told him. "The *SS Hauptquartier* in Herzogenrath was burned too and we're on our way to join the *Waffen SS* in Aachen."

Everything I said just came into my head. I was reacting not only to his questions but to the black uniform, the black hat, the silver trim, the insignia, the face, the vehicle, the gun on his hip, the intimidation, the violence of it all. This was the *SS*, the elite of Hitler's troops.

"You are going to join the *Waffen SS*?" he said.

"Ja!" I said briskly, just like he spoke.

He looked at me, then at Leo. If he said, 'Get in the Volkswagen,' we were done. He could pull out his sidearm and do it right here in the road, there was nothing to stop him.

His arm shot up in the air in the Nazi salute.

"Heil Hitler!"

I clicked the heels of my new shoes together and gave him the best looking Nazi salute he'd ever seen.

"Heil Hitler!" I shouted back.

The *SS* officer stepped toward me, shook my hand firmly, then shook Leo's hand. He was probably thinking Leo and I were going to make a pretty scrawny pair of soldiers but he wished us both good luck, then he got back in the Volkswagen and they all drove off.

When I gave him the Nazi salute and shouted *'Heil Hitler!'* an incredible feeling shot through me. It was like a victory greeting. Then when the Volkswagen drove down the road I felt for the first time in years—just for an instant—I felt I was free.

But the feeling was brief. It only lasted a moment.

We started walking again and almost immediately came to a crossroad. We turned right, walked a short ways and sat down. Leo was so scared he was shaking. I was scared too. At the moment the *SS* officer was asking me questions, I didn't have time to feel anything, I just talked to save my ass. But now I realized how close that had been, how very close. At this stage of the war if we got picked up, that was it. We weren't going to be sent back to the camp. We knew the attitude of the Germans. To them you were nothing, to shoot you was nothing. We had our new clothes but still, we had no papers and we were two young guys, hiking on the road and it wasn't a Sunday, so what the hell were we doing. I would have felt much less conspicuous walking down the main street in Aachen.

"How far do you think we are from the border?" Leo said.

"Maybe eight, ten kilometers," I told him. I was just guessing.

We started walking again. It was six or seven by now, full light, and at least we were heading in the right direction again. We just had to keep walking and hope we didn't run into anyone.

We walked for another hour or so when we saw something up ahead, some kind of construction. It was a tank trap. A good sign. I'd seen them before, on shopping trips into Aachen with Mama. Concrete pilings a good meter high set in the ground in rows like teeth, four deep, angled toward Holland. The Germans had built them all along the border so we must be getting close. This was the *Westwall*, the Siegfried Line, running down from the Baltic Sea all the way along the western boundary of Germany. When we got to the trap we stopped to look around. We might run into guards pretty soon. Beyond the traps were posters: *"Achtung: Minen."* "Attention: Mines." We couldn't go straight into the minefield and I was looking around trying to figure out which way to go when I saw two German sentries to our right, about a hundred meters away. I grabbed Leo and we ran behind a hedge and crouched down. The guards hadn't seen us. A couple minutes later they turned around and walked in the other direction.

"We've got to go that way," I pointed at the poster.

"Through the mines?" Leo said.

There wasn't any choice.

I ran through the tank trap, around the pilings, and into the minefield with Leo right behind. We ran like hell and either we were lucky enough not to step on any mines or they were for tanks instead of people because nothing went off and we made it. We ran for a good kilometer till we came to a field of clover and flopped down, exhausted. It wasn't good cover but we couldn't run any more. When I got my wind back I stuck my head up and looked around—there were the two sentries again! No, they couldn't be the same ones. We'd run too far. They were at least as far off as the first ones had been, walking towards us, and although it looked as if they were talking to each other this was a very bad spot to be in. Besides the guards, there were pill boxes along the strip in front of the tank traps.

"We can't get through here," I said to Leo. "We're going to have to go the other way, around Kerkrade."

We waited till the sentries turned around, then we took off again. It took more than an hour, going through the farmland, on the paths between fields, before I saw what I hoped was Kerkrade in the distance. Leo was drooping, dragging himself along. I was tired too, and hungry, very hungry. The sun was overhead, we'd been walking and running more than twelve hours. We started circling west again because I was sure we'd passed the end of the fence that blocked the road between Aachen and Holland. What we finally came to was a fence, but it was just two strands of barbed wire strung on posts with a sign: *"Achtung: Grenze."* "Attention: Border." There was no one around, no sentries, no tank traps or minefields or pill boxes. The fence went right through a field—half the farmer's field was in Germany, the other half in Holland. We stepped over the fence into Holland.

We were out of Germany but we weren't safe. If we ran into someone they might be more friendly but they could also be an *NSB*er, a Dutch collaborator. So we had to keep moving.

We started walking through the farm field. I could see a house in the distance, on the outskirts of Kerkrade. After a few minutes we came to a cobblestone street. It was the main road into town—and we could see the town now too, Kerkrade was only a kilometer or so away. We passed houses, built closer together as we approached the town, and for the first time since we'd escaped from the camp I felt, not inconspicuous, but not so out of place either. Then I saw something familiar. A trolley. Kerkrade was the end of the line, as far as you could go in Holland, and as we got closer I read the name on the trolley: "Heerlen." I looked back for Leo.

"Come on!" I shouted to him. "I'm going home on the trolley!"

He was so beat he hadn't even noticed it.

"We don't have any money, Josko!"

"Don't worry—just get on!"

We were almost there when the conductor clanged the bell. I ran the last few steps and jumped on, and Leo grabbed the railing and jumped up behind me. There was plenty of room, only five or six people on board, and we sat down on the wooden bench as the trolley got up to speed. It was such a relief to be sitting down, to be moving without using my legs. The conductor started at the other end taking fares and when he got to us he said—in Dutch—naturally, "Where are you going?"

"Wir gehen nach Heerlen," I answered in German.

"Vijf en tachtig cent," he said in Dutch.

"You dumb Dutchman!" I said to him. "We don't have to pay. We're Hitler Youth!"

The conductor was an old man and he gave me a look of great disgust, and just shrugged and walked away, as if he was relieved he didn't have to touch me.

No one else on the trolley paid any attention to us and we rode along, making the usual stops as the car filled up with people. Outside in the street the traffic was becoming heavy, mostly military—German trucks, jeeps, half-tracks, soldiers—but what I kept staring at were the ordinary people, the Dutch civilians getting on and off the streetcar, walking along the street, going in and out of shops and houses. Instead of all wearing the same ragged blue uniforms, they were dressed in their own clothes, and to see them behaving normally was so strange. We were in Holland and Holland was occupied by the German army, but unlike the camp the people weren't cowering and cringing, they didn't act apprehensive and hounded from being continually driven and beaten by guards. After two years in the camp, I'd forgotten what ordinary people looked like and it felt good to see them in the street, going about their business and sitting around us in the trolley. For a moment I was even relieved. We were on our way home and soon we would get something to eat. But I knew we couldn't stay in Frankstraat. We had no papers, and even if we did there were *NSB*ers around who might know who we were. We were still going to have to hide out till the *Amis* came.

Leo was so tired I don't think he could have made it if we hadn't caught the streetcar and ridden the last eight kilometers into Heerlen. When we pulled into the station in the center of town I had to give him a nudge. He hardly knew where he was, but being so close to home revived him. We crossed the square where everything looked just like the last time I'd seen it—the huge, ugly, orange 'Victory' banners the *Moffen* had put up were still there—but I was so hungry I barely looked around. All I was thinking about now was my stomach.

It was a good forty minute walk from the center of Heerlen to Frankstraat but we knew what was waiting for us. When we got there I sat down on a tree stump on the side of the road, across from where Frankstraat ended at the road into Heerlen, and waited while Leo went to his house. As it happened, his house was right on the corner, the first one on the end of the street, and I sat there watching while he went through the front gate, along the hedge, and disappeared when he went around to the back door. I heard him knock, and wait, another knock, then suddenly a young woman's voice shouted, "Leo! Leo! You're home!" It was Anna, his sister. I heard her shouting, "Mama! Mama!" and then his mother was shouting too. They all went inside and I heard the door shut.

I sat there looking up Frankstraat. It was empty, no one was out. The men were at work at the coal mines, the women were busy, the children at school. It was like everything else, familiar but strange. The houses with their peaked roofs built two and four together, all exactly the same, the hedges in front, the white pebblestone walks and painted curbs, the pocket parks with bushes and flowers still blooming in the middle of the wide, unpaved street. Five houses up on the right was #22, my own house, but I wasn't going home. The last time I walked down Frankstraat had been two years ago, on my way to the water tower, and walking down the street that day with nothing but the clothes I was wearing and thirty-five cents in my pocket, I had told myself I'd never come back here again unless I could walk up the street like a man.

"Josko! Josko!"

It was Anna, in the backyard waving at me.

"Come here, come inside!"

I ran across the road and around to the back door. Anna and Leo's mother gave me a big hug and pulled me into the house. She and Anna were crying, they kept saying they couldn't believe we were alive, and at the same time they were getting out all the food in the house and Leo and I were stuffing ourselves as fast as we could. They asked us all kinds of questions, but we were all so excited, so frantic, that no one made any sense. For the first time in two years Leo and I could eat, sit down at a table and eat as much as we wanted. Our stomachs couldn't hold it, but we tried.

"Where were you?" Mrs. Timmerman wanted to know. "How'd you get here?"

"We walked from Germany," I told her. "We escaped from a camp in Kohlscheid."

The town was too small, they hadn't heard of it.

"You were both in the same place? Is that where you've been all this time?"

"Yes."

"What was it like? Did they feed you? You're so thin, both of you."

"We're here," I said. "We got out."

"It was bad, wasn't it?" Leo's mother said. She was still crying, her arms wrapped around Leo while he was eating.

Leo didn't say anything. He couldn't describe what he'd been through. Even if he could, there wasn't time. We couldn't stay here, there were *NSB*ers living in Frankstraat.

Leo's mother went to get clothes for him while Anna made sandwiches for us to take. When Leo couldn't eat any more he put on some of his own clothes. They were just like the ones he'd stolen, a little short. He'd grown a couple inches in two years but hadn't broadened out any. His mother gave me some underwear but the clothes I had on fit so well I kept them. Anna tied up the sandwiches in a *kumpel,* a gray and white checked towel all the coal miners carried their food and clothes to work in, and we got ready to go. It seemed like we'd only been there five minutes.

"Are you going home, Josko?" Mrs. Timmerman asked.

I shook my head no. She didn't ask why. Everyone knew my situation at home.

"Gizela's still alive," Mrs. Tmmerman said. "She's still the same."

I hadn't asked—I'd been afraid to ask—about my sister. She had been beaten by German soldiers in Heerlen more than two years ago, just before I got caught. She couldn't walk and lay in bed all the time.

"Where're the *Amis*?" I asked her.

"Still in Belgium. We heard on the BBC last night."

After the Germans collected all the radios from everyone, Leo's father had built one himself. The BBC had a broadcast in Dutch and they listened to the news every night.

"Is anyone in the water tower? Are the *Moffen* watching it?"

"I don't think so," she said. "I heard someone was hiding there but I don't think the *NSB*ers know about them.

That's where we'd go. There wasn't much choice. It was all farms around Frankstraat and there weren't any forests or places to hide.

There was no one out on the street when Leo and I left, so we crossed the road, circled around through the fields and into the apple orchard to the water tower. When I climbed up the side and looked over the rim I saw the same sight as two years before, only this time there were even more guys hiding inside. They were older than the ones before too, mostly coal miners in their twenties and thirties. There was no one I knew but fortunately, just like the first time, someone recognized me. "Josko," he said. "Where have you been?"

"Germany," I said. "Working in a factory."

I didn't tell him we'd escaped from a camp because I didn't know who anyone else in there was. But it didn't matter. Nobody asked any questions. They all had their own problems too. Everyone was just waiting it out. The *Amis* were close but they weren't here yet, and if any one of us got picked up now, we were going to be shot.

Chapter 3
Water Tower near Frankstraat; Early September, 1944

The atmosphere in the tower wasn't as bleak as two years before because we all knew the momentum had turned and the war had to end soon—as soon as the *Amis* got here. But we also knew the Germans and the *NSB*ers were getting desperate and until the *Amis* arrived we were stuck. That anxiety plus the boredom of being cooped up in the tower and the problem of finding food made everyone edgy. During the day we climbed down in shifts of a few guys at a time to knock apples off the trees while someone in the tower stood lookout, and we also pulled beets up from a field near the orchard. But after a few days of nothing but apples and beets Leo and I had diarrhea, just like the others, so I said, "Let's go to Theo's farm and see if we can get something to eat." Theo's father owned the orchard and his farm was no more than a kilometer and a half from the tower. Theo had been in my class at school and he was the one who got beaten by the teacher for wearing an orange flower in his shirt on the Queen's birthday during the first year of the occupation, so I was sure of his father's sympathies.

In the evening at dusk we climbed down. Crossing the orchard and fields we didn't see anyone and when we got to the farm house and knocked, they let us right in. The family was eating supper with a couple of their farm workers and when

Theo's father saw me he grinned and said, "Josko! Where you been? Haven't seen you for a while."

"I've been away, working."

He knew what I meant.

"How'd they treat you?" he said.

"All right," I said. "I'm here."

"I saw some guys at the water tower," he said. "Is that where you're hiding?"

"Yeah, we're eating your apples," I said.

Everyone laughed. They had made a place for me and Leo at the table and gave us as much to eat as we could hold.

"How many are out there?" Theo's father asked.

"Twenty-five, besides us," I said. "They're pretty hungry too."

"We'll give you something to take back," he said. "We've got enough. The *Moffen* are too worried right now to be counting my potatoes."

We took a few loaves of bread back to the water tower for the others and Theo's father had told us to come back when we were hungry, so three days later Leo and I went back. We got the same reception, and brought food back to the tower again too. But the third time we went was different. As we approached the farm we saw a detachment of German soldiers parked around it. There were about twenty trucks and it looked like they were setting up camp, so we had to go back empty-handed and hungry. The next afternoon Theo's father came out to the tower to warn us the Huns were still there. "They've dug in," he told us. "It looks like they're staying." He said there were small German units all over the place now, some digging in for a fight, others getting their ass back to Germany. They were falling back as the *Amis* got closer.

We couldn't go back to the farm so the next day Leo told me he was going to go home again and get something to eat. We climbed down, cut through the orchard and circled around the fields, and got to Leo's without seeing anyone. It was like the first time, his mother was glad to see us, though not so surprised like before, and she gave us something to eat and tied up another *kumpel* full of food to take back. We didn't hang around. Things were more tense every day as the *Amis* got closer. Everyone knew something was going to happen soon and we had heard from some of the guys in the water tower that all the Germans in the area—the ones who'd lived in Frankstraat and Huiskens, another neighborhood full of coal miners—had gone back to Germany. Later I heard how emotional it had been, because some of the Germans, people I knew, had lived in Huiskens for twenty years, and after all that time they were leaving their homes. They were afraid of what might happen to them once things turned, even if they weren't the ones who had been nasty and arrogant.

When Leo and I came out the back door of his house, we looked around, didn't see anyone, and opened the small metal gate to the sidewalk. As we started to cross the road I noticed a guy coming down Frankstraat on a bicycle. At the same time two Dutch collaborators in their black uniforms with shotguns over their shoulders appeared. They were coming down the road toward us. They were still a good hundred meters away, a safe distance, but we'd intended to just cross the road, climb the low fence and head straight across the fields for the water tower which was close enough to see from Leo's house. Now we couldn't do that.

"Let's go the other way," I said to Leo.

We turned left on the road, away from the *NSB*ers, and walked toward one of the Dutch "Alps", the piles of stone tailings from one of the old mines. I figured we could circle around it and get into the field that way. We had almost gotten there when I saw two German soldiers coming up a path that crossed the road. They were about two hundred meters away, but if they turned left they'd be coming right toward us. And the *NSB*ers were still somewhere behind us. Then we heard a shot—what we thought was a shot—the guy on the bicycle had a blowout, and Leo and I dove into the potato field by the road. We crawled on our bellies till we came to a wheat field. The wheat was high, ready to be harvested, and we stood up, turned what we thought was away from the soldiers, and ran. The *kumpel* was gone, somewhere back in the road, and when we dashed out on a path, the soldiers were right there, no more than ten meters away, almost as if they were waiting for us. They were *Wehrmacht*, wearing the green uniforms of the regular army, with their

Mausers shoulder high, not aiming but ready. I stopped, froze. If they were young punks they probably would have shot us right there but they were older veterans, very cool. They wore heavy boots, their uniforms were dirty, and instead of steel helmets they wore caps since this wasn't a combat area. Not yet anyway. One of the soldiers motioned me over.

"What do we do now?" Leo asked me.

"Ik weet het niet," I said.

I didn't know what to do but I had my hands up and as we walked toward the soldiers, the two Dutch collaborators came around the corner of the wheat field, running like hell, saw us on the path, stopped and turned and ran toward us. The soldiers didn't say anything, they waited for the *NSB*ers. I recognized one of them right away, an older man, close to sixty, who lived in Frankstraat. When they got to where we were standing, he looked right at me but didn't recognize me. I was almost seventeen and it had been awhile since he'd seen me. He said, "Show me your papers." Speaking in Dutch, of course.

"I don't have them," I said. I took a belligerent attitude toward him immediately.

"Why not?"

"I left them at home."

Now the German soldier said, *"Wo ist dein Ausweis?"*

Same question, where are your papers?

I replied in very smart German:

"Ich habe keinen Ausweis." I have no papers.

"Why don't you have any papers?" the German said. "Are you a saboteur?"

"We are not saboteurs," I answered. "We are prisoners. I was in a camp—*Ich bin aus dem lager gefluchtet*—I fled from a camp, I've made my way to Holland and now I'm hiding. If you turn me over to those people they will shoot me. *Wenn Sie mich in das Deutsches Haus bringen, werden die uns auch erschiessen.* And if you turn us over to the Deutsches Haus, they will also shoot us."

I had nothing, no papers, nothing to tell him but the truth, and I was putting myself in his hands. But I told this to the German soldier, not the collaborators. The soldiers were haggard looking in their boots and dirty uniforms, and looked like common, honest men who had seen it all. The one I was talking to was a sergeant and I addressed him respectfully with his rank, *Herr Feldwebel.* Although he was an infantry soldier he looked out of place, more of a studious man, and he was smoking a bent, Sherlock Holmes pipe. He stood there looking at me, then he took the pipe out of his mouth and said:

"Do you think the Americans are going to win the war?"

I said, "Yes, I believe they will win the war."

He said, "Do you know what's going to happen if they do? There's going to be unemployment and inflation, it's going to be another depression, just like 1930 again."

I didn't say anything. I had no answer for him—I had my own depression to worry about—but he was very serious. He had thought about these things, in the middle of the war.

While we were having this conversation in German, the Dutch collaborators stood there. I don't know how much German they understood, but what I did know was if the German soldiers turned us over to them we were done. They weren't real soldiers, they still had to prove their manhood so to speak, and because of the way they had treated people, if the *Amis* got here they would have a lot to worry about.

"What are you going to do with us?" I asked the German soldier. "Are you going to take us in?"

He looked at his companion and said, "What do you think? What should we do—should we shoot them?"

The other soldier said, *"Nein, lass die Flegel gehen..."*

No, let the punks go, he was saying.

The sergeant said, "What about these Dutchmen?"

"We can take care of them," the other soldier said.

The sergeant shoved his *Mauser* in the Dutchman's stomach, looked at me and said, "If I see you again, I personally will shoot you. Now—*Verschwinde* your ass!"

Disappear your ass!

Leo and I took off through the wheat field and for the first time he was ahead of me. We ran all the way to the orchard and climbed back up in the water tower.

We had been hiding out more than two weeks and except for the two trips to Theo's farm and the quick meal at Leo's before we lost the *kumpel,* it had been nothing but apples and beets. With all of us stuck up in the water tower, the atmosphere was tense. There was nothing to do, guys were sick from the diet and also hungry, very hungry. We were just waiting, trying to hold out, everyone afraid of what might happen before the *Amis* got there.

It had been several days since Theo's father came out to warn us, and I said to Leo, "I'm going to try to get to the farm tonight." I knew it was risky with all the Huns camped near the farm but my hunger dictated getting something to eat. "If the *Moffen* are still there, I think we can sneak past them," I told Leo. "They've got other things on their mind besides a couple of punk kids."

We waited till it was good and dark before we climbed down. When I saw the silhouettes of the farm buildings in the distance and no sign of the Huns I thought they'd left, but when we got closer I saw the trucks were still there. The soldiers had moved them to keep out of sight and now they were parked in a gully between the dirt road and the farm. We crept through the field up to the edge of the dirt road and stopped. It was quiet. We waited. It was a big farm, two barns by the house with cows and pigs, and another building with almost medieval quarters for the workers. I heard noises, somebody walking, heavy boots dragging along on the packed dirt. Two soldiers walked by—sentries on guard duty. We stayed still. A few minutes later they came back from the other direction. They were patrolling the road, back and forth, passing a few meters in front of us. I suddenly realized how dangerous this was because the only way to get to the farm was to go down this dirt road, right past the trucks. But we were here now, and I was so hungry my stomach said, go, do it. "The next time the sentries go by," I whispered to Leo. "Then we'll go."

When they were a good fifty meters beyond us I moved. As dark as it was, I could easily see the outlines of all the trucks parked down in the slope of the gully and I figured the Huns were all camped on the other side. We tried to walk fast, but bent over, quiet. Up ahead was the farm gate, no more than fifty meters away, and beyond that the pit where they dumped all the manure from the animals. Leo was right beside me. We got past the first truck. No sign of activity. We kept going.

"Was gibt es?"

A *Mof.* We stopped. The voice came from almost beside us.

"Who's out there?" Another voice, behind us. "I heard something. Did you see anyone?"

"No." That was the first sentry. He yelled back, "I didn't see anything!"

But the sentry behind us was moving toward us. I could hear his boots on the road and saw someone else moving by the trucks, so we got off the road, back into the field on the other side. As the sentries approached they shouted back and forth to the soldiers down by the trucks. No one was very excited, they were just asking if anyone heard anything. Leo and I kept moving. We got fifty or sixty meters out in the field and stopped to listen again. We could still hear the commotion up on the road and I was trying to figure out what was going on too. I was confused just like the Germans because I was sure they hadn't heard us. If we'd made any noise I would have heard it myself—and the first thing I'd heard was the *Mof* asking, *"Was gibt es?"* By now it sounded like ten or fifteen Huns were calling back and forth. Then someone shouted an order: "Spread out and check the area!"

We took off. We ran through the field without caring how much noise we made because the *Moffen* were making such a racket themselves. We had only run a short distance when I heard a shot, a rifle, and from the sound I knew right away someone got hit—not close to us, but not that far either. We ran away from the sound, running like hell now. Voices were screaming, shouting orders in German. We ran out of the field to a path and stopped. The path ran between two rows of poplars. Now we were exposed. I ran back in the underbrush between the path and the field, tripped and fell down. I started to get up when Leo fell on top of me. Now we were the ones making a racket and although the voices were some distance away, back toward the farm, what if they'd posted another sentry further away? I scrambled back up to the path—at least now I could see where I was running—and as I ran I pulled out a knife I'd gotten at Leo's house, a short file ground down to a sharp point, set in a wooden handle.

After all the frantic stopping and starting and zigzagging to get away from the commotion, I felt a pain in my side. I had to rest. I ducked off the path behind a big tree—and bumped right into someone. One of the sentries! He grabbed me. I swung the knife and stabbed it into him hard, as hard as I could. He let out a fierce yell and let go. Someone else swiped at me, punched my shoulder, and almost knocked me down. I got my balance, scrambled back to the path, and ran. Ahead of me was someone else, running. It was Leo. He looked around, surprised to see me suddenly behind him instead of in front. When I caught up he said, "What happened? I heard someone scream."

"I stabbed a guy," I said.

We were panting, both out of breath. We ran off the path into the field. Up ahead was the orchard. Now we could walk.

"Who was it? A *Mof*?"

"I don't think so," I said. "I don't know who it was. But I hurt him. I hurt him bad."

The next day I was out picking apples when I saw someone in the field pulling beets. It was one of the *knechts,* a guy who worked at the farm for Theo's father. There was no sign of any *Moffen*, so I went over to speak to him. As I walked up he recognized me and said, "Stay away from the farm."

"What's wrong?" I asked him.

"Last night they sent a guy out with food for you guys and the *Moffen* shot him. Another guy got it too. He was walking ahead—two guys went ahead of the guy carrying food—and they ran into someone and one guy got stabbed."

I went back to the orchard. The *knecht* was dead. I'd known right away when I stabbed him it wasn't a soldier. After I tripped and fell, I got out the knife to protect myself, in case there were sentries posted out further, and when the guy grabbed me—he probably thought I was a *Mof* too—I just reacted on instinct. By the time I realized it wasn't a sentry—there was no leather belt, no buckles or helmet—it was too late.

I sat under a tree and watched the sun set.

Once before, in the labor camp, I had killed someone. But that was different. It was the first year, when they were training us to be *Steigers* and I was working in the shop building lorries. There was a guy in the shop, a German guy, very strong, very arrogant, who was always talking about how he was going to quit and go join the *SS*. He ran the rivet gun and when he finished riveting a lorry he'd ram the gun against the side so it made a terrific noise, BRRRRRR, inside the lorry where someone like me was holding the other side of the sheet metal. He did it just to harass you, make you nervous. He'd been an athlete and would grab people and show off his strength. One day he got me in back of the shop and...he raped me. I fought him but he was too strong. All muscles. He had done this before to other people. It was just a question of who was going to be next. Three days later when I was working in the shop they sent me out to the yard in the complex to get wood. The complex was huge, there were always people doing different jobs, moving around the yard, and out where they kept the lumber there were logs stacked up. Some of the Russian POW's had made hideaways in the piles of logs, places where they could hide out and rest, get away from the guards, maybe make out with one of the Russian women because there were women in the camp, even working down in the coal mine. The Russians also carved things in the hideaways, figures of birds on wheels and when you pulled them along the wings flapped. They traded them to the German coal miners for food and sometimes I saw one of the Germans leaving the complex at night with a carved bird under his arm. At one end of the log piles the logs were stacked vertically so it created a kind of lean-to and as I approached the log piles I saw what I was looking for. I had seen the athlete leave the shop and I knew he was somewhere out there too, prowling around the yard. I knew what he was doing. He was looking for someone else to grab, maybe one of the Russian women, and when I saw him duck inside one of the hideaways I picked up a rock and followed him in. As I came in he turned and saw me. He recognized who I was but I had already hit him. As he went down I kept hitting him. I crushed his head with the rock. It was the first man I killed, and it was the only time I didn't feel bad afterwards.

When I came out of the hideaway I didn't bother to look, I didn't care if anyone saw me. The reputation the guy had, no one would have been surprised what happened to him, even the other Germans. Anyone might have killed him. One of the Russian POW's, for raping their women. I picked up some wood and on my way back to the shop in the main building, by chance I went past *Herr* Schmidt. When he saw me he gave me a big *"Heil Hitler!"* like he was glad to see me working hard. I saluted him back, gave him a wave, thinking he wouldn't be smiling if he knew what I'd done to one of his fellow Huns. I didn't go straight back to the shop. Instead I went to the library. The labor camp had a library for the coal miners and the other Germans who worked at the complex. I went up to the window which was separated from the yard by a fence, and said to the librarian, *"Haben Sie Bücher von Zane Grey?"* He was a little old German, a former coal miner, and seemed surprised at my question. "You read Zane Grey?" I said yes, and asked him again if he had any books by that author. He smiled and nodded, but I don't know if it meant he had any or not because he wouldn't give me a book. He told me they weren't for the use of the prisoners.

My friend Petar had introduced me to Zane Grey, before the war. Petar was older than me, also from Yugoslavia, and reading the books in Dutch and German we talked about how we were going to America to be cowboys. We were just a pair of ignorant kids who knew nothing about America, but to us the "wild west" was such an exciting life. It was freedom. It was away from the life of the coal miners, the hard work, the drudgery, the danger underground, and away from the life of the guestworkers where everyone lived with the fear of losing their job and being sent back home. There was also the frustration of living in a place that was not really your home, where you were always reminded you didn't belong. So for Petar and me, the books, just the unusual sound of the name "Zane Grey" was special. It was also important that Petar was the one who showed me Zane Grey because when I first came to Holland Petar had taken me under his wing, like an older brother. He was someone I could speak to before I learned Dutch, and although I learned I could take care of myself, if there was trouble Petar was always someone I could count on. He was a very angry, determined young guy, the kind who got in trouble because he wouldn't accept living on the fringe the way we did, the constant fear and anxiety that you didn't belong where you were. One day in the school yard he fought two Dutch kids and beat them so badly it made me feel sick, as much as I wanted him to defeat them. As it turned out Petar's father did lose his job in the coal mine and his family had to go back to Yugoslavia. After the truck took their few belongings away, they spent their last night with us, then they got on the train, said goodbye, and I never heard from him again.

After the war began I always imagined Petar fighting with Tito and his partisans against the Germans, up in the mountains. That had been my dream too, to fight the Nazis with Petar and Tito. Tito was my hero and Petar was my friend, and I had always lived with the hope I could go back to Yugoslavia, especially before the war. All my good memories were from when I was living on my grandfather's farm in Vrbosko, although it seemed like a long time ago and my memories were already faint.

Chapter 4
Vrbosko, Yugoslavia; 1934

My grandfather, Big George Filipovic, had a farm outside the town of Vrbosko where he raised horses. He had been very successful, one of the better off men in the neighborhood, but when he supplied horses to the Austro-Hungarian army during World War 1 it turned out to be the wrong side. After the war most of his land and all his horses were taken from him. Big George was left with only a few acres where he could raise some crops and keep four cows. They had to divide the house in two and the family shared it with the postmaster.

There were four of us in the house; besides myself and my grandparents, my Uncle Pepi lived with us. He was an engineer and had no wife or children of his own so he spent a lot of time playing with me. It was a big loss when he got a ruptured appendix and died suddenly. My grandfather had died by then too, and soon afterward my grandmother said, "Josko, your father is going to come and get you and take you to Holland to live."

I had never seen my father and I only knew about him because of the packages of food with things like marmalade and canned fish sent from Holland. He also sent money, a few dinars my grandmother stacked on the table by the package of food. My father came by himself and we took the long train ride to Holland, waiting on platforms, changing trains several times, before we arrived in the middle of the night in Heerlen. We took a taxi to the house where I met Mama and was taken in to meet my sisters, Gizela and Dragitsa, who were glad to see me although it didn't really register that they were my sisters. At first they were just older girls in the house I was now going to live in. The next day Mama came outside to throw out a pan of dishwater, then she called me over and said, "Come and give me a hug." It was strange because I'd never had a mother before—I had my grandmother Oma and Big George and Pepi—so I never missed having a mother. But it was a good feeling hugging her.

The family lived in Heerlerheide, outside Heerlen, in an apartment on the second floor of a big hall used by various ethnic groups for meetings and events, and there were three young men, also coal miners from Yugoslavia, boarding there too. Only a few months later we moved to Frankstraat which was a step up, a new development built for the guestworkers and coal miners to live in.

Frankstraat had about seventy two-story houses, all built the same, either two or four joined together. There were three small rooms upstairs, one for my father and Mama, one for the girls, and the very small one was mine. Downstairs were also three rooms, the formal living room with a door that was always kept closed except for three or four times a year, like at Christmas, and the kitchen where we also ate. The washroom, the laundry, had a big, round wooden washing machine. There was a front hall and a cellar which was only a few cement steps down to a space to keep potatoes. In back was a small garden with a cage where my father kept a few rabbits and another with his homing pigeons, and in front all the houses had a neat Dutch hedge by the street. The only running water was cold and the bathroom was like an outhouse only inside and you flushed it with a bucket of water. We heated the stove in the kitchen with either coal or *schlamm,* a kind of mud made of coal dust that had settled in water after they washed the coal and stone at the mine. In the winter it froze and I chopped it with a hatchet in the back yard.

I had a little trouble at school at first because I couldn't speak the language but I began picking it up. One day I started waving my arm in class and the teacher finally called on me, just to shut me up, and I stood up and called out the names on the blackboard as he pointed at them. *"Appel, roos, vuist, neus..."* It was all sound, I didn't know what the words meant, but I had a good ear and when the teacher said, "Fine, you can sit down," I went up and took the pointer from him and recited the Dutch words again: apple, rose, fist, nose. I had my chance and didn't want to stop! It also took a while to establish myself with the other kids who were always teasing me, pushing me, knocking me down. Then one day when they formed a circle around us, I knocked down the Dutch kid who was pushing me and it was so easy I began throwing him around just for the fun of it. After that we became friends and I went over to his house to play, although play mostly consisted of loading coal on his father's truck.

My father worked in Orange 3, one of the four local coal mines which were all named for the royal Dutch House of Orange, and it was about two kilometers from Frankstraat. The men worked in shifts, constantly changing from day to noon to the night shift. Everyone knew it was very hard work. Also that it was dangerous. Losing fingers was common. Getting killed was not that unusual, not from explosions because they were very careful about gas, but from collapsing tunnels. The miners got so busy digging coal that they didn't stop to shore up the tunnel behind them. They got paid for the amount of coal they dug, not for putting up shoring timbers, so the miner was always thinking he could chop just another meter or two before he stopped to put up a timber. They were paid a basic wage for their five and a half day week, but if you chopped more coal than your quota you earned a premium. Or if you worked overtime, which the miners all loved because that was the only way to get out of the rut and make extra money. The more you produced the more you were paid, that was the incentive. My father got paid on Friday and his pay receipt was a long paper strip, as long as your arm, with his base earnings, then the premiums for extra coal on top of his wage, then a deduction for rent, for electricity, for water, wood, coal, and *schlamm*. Mama always looked at the pay receipt carefully to make sure it was right. The next day, Saturday, I was sent to the store back in Heerlerheide where the family had credit and I paid last week's bill. If a few cents were left over it was applied to the groceries I picked up that we owed for this week. Like everyone else, we were always a week behind.

Our neighborhood was almost all foreigners. Besides other Yugoslavians like Petar's family, there were Poles, Hungarians, Italians, Germans, and a few Dutch like the Timmermans. Most of them had done the same thing as my father, they came and worked and made enough money to bring their family. I think every guestworker had the same dream, to save enough while he was working and then go back, buy a piece of land, a house, and live a comfortable life in his own country. The average coal miner earned between 25 and 35 Dutch *guldens* a week and five *guldens* might be equivalent to a civil servant's salary for a week in Yugoslavia. So if you could manage to save five *guldens* a week—though it was very tough—and do that for twenty years, maybe you'd have enough. You also got a pension after twenty years. But that was where the manipulation came in. Because for a coal miner to stay healthy for twenty years was very unusual, and if you got sick you got laid off and shipped back home. All the foreign laborers were younger men anyway, in their twenties and thirties. They had agents to recruit foreign workers and they didn't even take older men.

It was an atmosphere where money was on everyone's mind and you had to scrape to save anything. Mama made all of Gizela's and Dragitsa's clothes and she also made clothes for the big Dutch woman across the street who had four sons, all coal miners living at home, so she had plenty of money. When my father needed a new pair of shoes, Mama put on his old, worn out ones, rode her bike over to Aachen in Germany, bought a new pair and put them on so she wouldn't have to pay duty at the border. No way they ever fit her, but the guards didn't look at her feet. She and my father also got their teeth in Aachen. A set of false teeth cost about 40 *guldens*, and after deductions my father was left with 17 or 18 *guldens* a week so it was a lot of money to spend. One afternoon Mama was sitting in the kitchen peeling apples and scraping the peels into her apron. When she got done she got up and threw all the peels in the fire. As soon as she did she let out a shriek because she realized she'd taken out her teeth and laid them in the apron too. She grabbed for the poker but by the time she pulled her teeth out of the stove it was too late, the heat had already melted them.

At home we spoke Serbo-Croatian, the Yugoslavian language, and I picked up Dutch rather quickly because I had a good ear. I also learned enough Polish to speak to my friend Vwadzu Kamic's parents, and later I learned German and English in school. My father spoke quite good Dutch and German just from hearing it around him—he also had a good ear—but he couldn't read. Even in his own language he had to spell out each word in a newspaper like a child.

You heard all kinds of languages in the neighborhood, although we spoke mostly Dutch on the street, and I had a lot of Dutch friends, like Leo and especially his brother Scheng, and Joobi and Theo. There were some German guestworkers too, like Wil who lived next door with his Dutch wife, and my friend Tutschki who was German despite his name. He was the daredevil who jumped from roof to roof on the houses and always tested the ice down at the cesspool at the bottom of Frankstraat. All the shit from the street ran down to a concrete reservoir at the edge of the farmland and the farmers scooped it out to fertilize their land. Whenever the frost set in the kids would go ice skating on it. But first somebody had to test the ice. Who did it? Tutschki. He'd let himself down the side of the reservoir, slide out on the ice, we'd hear a CRACK!...CRACK!...CRACK! and Tutschki would disappear through the ice. He'd come up covered with shit and grime, and on his way home he'd sing a song about how he fell in the cesspool. When he got home his father made him take off his clothes in the yard, zero degrees Centigrade, and he'd hose him off, all the time cursing him. Two days later, who goes down to check the ice? Tutschki! He was the hero, the one without fear. Tutschki was crazy but he wasn't stupid. He did well in school and everyone liked him, so it was sad when his father lost his job and the family was sent back to Germany. It was also bad luck for Tutschki because later during the war he was forced to join the German army, and we heard that he was one of the first soldiers killed when they invaded Holland.

There was only one kid on the street whose father wasn't a coal miner. Jan te Poel's father was a bureaucrat, a white-collar worker, and Jan was a skinny, blond, blue-eyed kid who just couldn't cope with these mean Dutch, Polacks, and Yugoslavs. Jan was in my class in school so he decided to side with me, and I walked him home every day for a while to protect him. As a result we became very good friends, which was why it was so disappointing to me later during the war when Jan joined the Dutch Hitler Youth. Perhaps it was the appeal of the uniform with the knife as a sidearm for the skinny kid who'd always been pushed around and couldn't take care of himself. Whatever the reason, it was a great disappointment to me.

For the first year and a half my life in Holland was like the other guestworkers' kids. It wasn't easy but I ate, went to school, did the chores, and had my friends. I was like everyone else, except for one thing. That was my name. I didn't even notice it till one day Gizela and Dragitsa and I were picking blackberries on the mountains of stone called the Dutch "Alps" and some of the other kids were teasing them because I didn't have the same name. They were Merkas, I was Filipowic, and for some reason that meant I wasn't their real brother. The other kids called me a name, a word I had heard but didn't know what it was, and now I realized it was something shameful. That was how I learned I was different and although I didn't know why, I realized there was something very embarrassing about it to my sisters.

It was when my father got sick that things began changing. I wasn't aware of the reason things changed. I was too young. All I knew was that something was wrong. It started when he had trouble breathing. The most common ailment among the coal miners was what the Dutch called "stone lung," where the dust settles in the lungs. When he felt a constriction in his chest he assumed that's what it was, but when he went to the doctor they found that he had an enlarged heart. In September, just after I began second grade, he went to the Catholic hospital in Heerlen where they cut him open and operated on his heart. He was in the hospital a couple weeks, then home recovering for several weeks after that. I don't know why the coal mine didn't just lay him off and send us all back to Yugoslavia because I'd already seen it happen, like to Petar's family, and Tutschki's, and I heard my father talking about it, worried they might do the same thing to him too. But at the time there had been some very active political campaigning by three parties trying to get people's support, especially the coal miners. The parties were instrumental in defending people's causes and on the basis of that they got their membership. My father had joined the Red Party, the communist one, and every week the man came and collected the dues—not much, maybe fifteen cents—and put a stamp in his book. When you had a complaint, if you'd kept up your dues, you could go to the party and they would fight for you. So the Red Party may have helped my father get medical treatment instead of sending us back.

Nevertheless, he never knew what would happen and besides the fear of losing his job, the illness was a very hard blow to his pride. Back in Yugoslavia he had been trained as a tool and die maker but when he couldn't find work in his trade and came to Holland, his attitude, like a lot of the guestworkers, was that he would show these lazy Dutchmen what a day's work was. He was proud that he always chopped more coal than anyone else, and everyone knew how much coal you dug because it was right there on your pay strip every week. It wasn't how much you said you dug. It was printed in black and white. In the situation he was in, that was all he had to prove what kind of man he was. His pride was not only how hard he could work, he was also very conscious of his appearance. He paid an Italian tailor to make him a suit and when he put it on he twisted around to look at the shoulder, and when he saw a small wrinkle he made the tailor take off the sleeve and fix it. It had to be perfect. On holidays and Sundays when he wasn't working, he'd put on his suit and walk up and down the street so people could see how handsome he looked. He was good-looking and well dressed and although he was a common coal miner, he wanted to set himself apart from what he was.

While he was home recovering from his illness and the operation, he and Mama had a big fight. Sometimes they argued, but this was different. It was a Sunday morning and when my sisters and I came home from church what began as shouting became physical. My father was ranting, Mama was screaming, then they started slapping each other, punching it out. Gizela and Dragitsa were crying and Mama grabbed them and they all fled into the formal living room and slammed the door. My father went upstairs, got a suitcase from under the bed, and forced open the locks—for some reason it was locked, although there was nothing inside—and then he stood there cursing and raving over the empty suitcase. I had stayed with him when Mama fled into the living room and I was very upset, very confused. I had no idea what he was doing, and if his intention had been to leave—where would he go? But I wasn't thinking about that at the time. All I knew was that he was so angry, raving and cursing.

After his big fight with Mama things began to change for me. Now for some reason things at home weren't right, they weren't normal. He started cursing me and soon it was nearly every day that he would say, "Josko, get the *schlauch*." I had to go into the laundry room and pull the hose off the washing machine, a rubber hose almost a meter long, and bring it to him so he could beat me. He would beat me for any reason. Or for no reason. It made me so angry to be beaten for nothing that I wouldn't cry, I couldn't cry. I couldn't give him that satisfaction. It made him even more angry when I resisted him, so he'd double the hose. Finally, when the pain became so great, I whimpered and cried, and he stopped. So I learned if I cried he would eventually stop, but by doing that he had not only beaten me but degraded me too.

My friends got a beating if they did something wrong or got in trouble, but then it was over and you were the son again. This was different. I kept hoping things would go back to the way it was before, but instead it got worse. I was getting it all the time, not with the hand like before when I did something wrong but with the *schlauch,* and always there was the cursing, as if for some reason he hated me. As boys we wore short pants so you could see the welts on my legs when I went to school or to my friends' houses, and sometimes the people I knew well, like

Vwadzu's or Scheng's mother would say something about it, ask me if I was all right. I said yes, I was all right, and told them not to say anything because if my father thought I was talking about him outside the house, it would set him off even more.

One night at home after supper there was a knock at the door. My father said, "Josko, see who it is." I went into the front hall, opened the door, and standing there were three tall Dutchmen, very formal in their black suits. They said they were a delegation from the town, and one of them said, "Are you Josef Filipowic?"

I said yes.

He said, "We have heard you are being beaten. Is that so?"

I had a feeling of panic. If they saw the welts on my legs, they would know it was true, and if they confronted my father, he would think I had told someone—reported him to the Dutch officials—and then he would go out of his mind with anger. Already he lived in fear of losing his job, and to get in trouble with the Dutch authorities meant he'd be sent back—all of us would be—back to Yugoslavia.

So without saying anything I closed the door.

The delegation of Dutchmen might have meant well, and someone obviously was trying to help me, but the reality for me was—I was on my own. This is what I had to live with. So I had to make a decision.

This was also the time I began to be so aware of food. Before things changed, I used to eat what the rest of the family ate. Now my meals were different. In the morning I got up before anyone else and started the fire in the stove, then Mama came down and made coffee and breakfast. The others ate cereal or eggs but my breakfast was the scraps of bread left over from the day before, whatever Gizela and Dragitsa hadn't eaten. The bread would have some rancid butter on it and I'd break it up in my coffee and eat it with a spoon from an enamel cup. Dinner I shared, though I always got less than the others, and as a result of my diet I was constantly hungry. I became conscious of food all the time. On Sundays if Mama cooked a chicken, I got the wings. After dinner, Mama would put the rest of the chicken in the cupboard and if it was a nice day the family would go out for a walk, visit the neighbors, go see the Milicic's in Heerlerheide, or do something special like go to the circus. I never went with them. Instead, I was left home "to watch the house." That was the excuse for leaving me home. To make sure no one came in and stole something. There was nothing to steal in the house. But there was a piece of chicken. I had seen Mama put the plate with the leftover chicken in the cupboard, and because it was so appetizing and I was so hungry I got carried away. They had left a lion with a gazelle—and the lion ate the gazelle! It was not only my hunger, it was my rebellion. But it wasn't worth it because then I had to wait for them to come home, and I knew what would happen. The first thing they did when they came in was go to the cupboard, open the door, and take down the plate with nothing left but a little smear of chicken fat. I think they must have enjoyed it because they knew what would happen if they left the lion alone with the gazelle. It happened every time, and I got it from everyone, from Mama and from my sisters, and then my father would say, "Josko, get the *schlauch!"*

My neighbors also knew I was hungry. My friend Vwadzu stopped on his way to school to pick me up so he saw what I ate for breakfast, and when I went to the Kamics' house his mother always gave me something to eat. One Sunday afternoon when I was outside in the yard "watching the house" while the family went out, I heard a voice on the other side of the hedge.

"Josko!"

It was Sonya, the young woman who lived in the house next door.

"I've got something for you."

I saw her hand come through the lower part of the hedge with a sandwich.

"Take it," she said.

I took it. The sandwich had a meat like salami. I never got meat like that and it was delicious, but after I ate it I worried the family would find out. If my father knew the neighbors were feeding me he'd go into a rage, saying I had embarrassed him by begging for food. I never said anything to Sonya but she knew the family left me alone when they went out because after I cleaned up the house I always came out in the yard and if my friends went by I talked to them through the hedge. After the first time she often brought me something to eat when she saw me alone in the yard, and although I was always afraid my father would find out, I always ate it.

As things at home became worse, Wil and his wife, who lived in the other half of the house, took more interest in me. They were young, perhaps thirty, with no children of their own, and Wil often tried to get me out of the house. He would say he needed my help doing something, and although it didn't always work when he asked, sometimes my father let me go. Or if my father was working the night shift, Mama usually felt safe enough to let me go over to #24. While I was there, his wife always gave me something to eat and Wil and I played chess. He had made his own chess pieces and had a thick book with diagrams and theories of how to play, and after he taught me the game we'd read the book together. Although Wil had never gone beyond 8th grade, he was very bright and people came not just from Heerlen but also from places like Leiden and Hilversum to play against him. In Germany he had trained as an automobile mechanic and ended up in the coal mines because he couldn't find work in his trade. I think he must have felt the same frustration as my father because he hated the life of a common coal miner and chess was his outlet. It became an outlet for me too, a way to get out of the house.

School was also an outlet. I loved school because I had my freedom. My academics were good but I was pretty disruptive. I got a 10 in math and German—the highest grade—but in behavior I got a 4 which was the equivalent of an idiot. I had a good relationship with my teachers but sometimes that backfired. One day in *Mijnheer* Van Dijk's class we were doing style-writing and Vwadzu hit my arm while I was writing so the pen splattered ink all over the paper. It was almost four o'clock, time to bring the papers up to the teacher and when I handed mine up to *Mijnheer* Van Dijk sitting on his podium, he looked at it all smeared with ink and got a look of disgust, as if it was an insult to give him such a mess. He swung his hand at me with a big ring of keys he was holding, I ducked and hit my head on the corner of his desk. He picked me up from the floor, took me to the basin where he put a coin, a big *gulden*, under the cold water tap and put it on my head to stop the swelling. He felt so bad he said, "Josko, I'm going to take you home." I told him, no, please don't do that. He insisted. When we got there, he waited outside while I went in and told Mama the teacher wanted to speak to her. "Bring him in," she said. I brought *Mijnheer* Van Dijk in and while he was apologizing my father came home from the coal mine. The teacher began explaining to him what had happened and my father went crazy, he beat the hell out of me right in front of him. After that, the teacher understood what I meant when I said, 'Don't do me any favors.' He and I became friends; he knew I had a tough time at home.

Church was also an escape because once again it was a chance to get out of the house. As long as there was a mass, a novena, or a First Friday, I could always say, "I'm going to church," and no one at home would question it. On Sundays they took attendance for the children so we had to go. Mama and my father never went but they gave me and my sisters a penny, a cent, for the offering. Of course I never put mine in. But just to be in the church was such a relief. St. Anthony's of Padua was a beautiful church with annexes so if you faced the altar there was a separate place on either side for boys and girls. In May there was a novena for Mary, the Holy Mother, we had a procession, and I was chosen to hand out novena prayer books which was a very big deal. I stood on the women's side of the church where I saw all the girls, then went over to the men's side, but I had no interest in giving books to the boys or the men! At novenas we all sat in the big church so they couldn't take attendance and unless I was chosen for something special, I skipped out with my friends and raised hell. These were great liberties and I loved it because I had my freedom. One Sunday in May when we were supposed to be in church we went out looking for birds' nests, to see if the chicks had hatched from the shells. One of the older Dutch kids ripped out a nest I had found and when he showed it to me with the beautiful blue eggs all busted—he had destroyed them, for nothing—I grabbed him and we tangled ass. He shoved me down on the path and dragged me through the mud between the hedges, and although I got beat—and had another treat waiting at home—I thought it was one of the good fights I had.

The other escape I had was taking walks with the old priest, Father Nicholas, who had started coming to the house. No one at home could challenge the church's authority, so if the priest said Josko was going out for a walk, then Josko went out for a walk. Like all the priests at the school, Father Nicholas was a Franciscan. He wore a brown robe and had the bald spot on top of his head where they shaved their hair. He was a big stately man, a scholar, and had a very tolerant attitude toward religion. He said other people were sincere in what they believed too and although he stopped short of saying you should be anything but Catholic, he didn't say you had to be a member of the Catholic church to go to heaven. We discussed all kinds of things—mostly him talking and me listening—and naturally when the other kids saw me walking down Frankstraat with the priest they wanted to join in. So there would usually be a crowd of us walking along while he talked and talked—then suddenly he would stop short in the street and say, "What did I just say?" The kids who were laughing and talking would stop, there was silence, then I answered his question. He scolded the other kids and said, "See? Josko is the only one who listens!" After that I listened even more! But I *had* to listen to him because he was my relief, my outlet, the one who got me out of the house. He started coming during the summer after school was out, two or three times a week, and continued into the fall when he started coming every day, talking to me about poetry, literature, religion and philosophy, and in the meantime my whole world was falling apart. That fall my father had a second operation. Whatever they'd done before, it wasn't enough, and he had to go back to the Catholic hospital in the center of Heerlen. They had to open up the cavity again and make room for his heart because it was so enlarged.

After the first operation the year before, they had let him go back down and chop coal, but when he got sick again that was over. Now they gave him a job taking care of the horses they used in the mines to pull heavy loads. They kept the horses down below so long that when they couldn't work anymore and brought them up, the horses were blind. Working in the stables was a steady job and paid a steady wage but there was no premium, no way to earn extra, so to my father it was an invalid's job and in his own eyes that's what he had become, an invalid, tending horses. It was degrading and I knew it hurt him because I couldn't help hearing him complain about it at home. He was no longer the big stud chopping more coal than anyone else—now he couldn't chop coal at all—and as bad as things had been at home, they got worse. After the surgery that fall he was home several weeks recovering, and that was what I dreaded the most, the times he was home, because then he had the most time to beat me. In early December, the sixth, something happened before he went to work that set him off. He began with the usual cursing and raving, and picked up a wooden ladle on the table and started hitting me on the head with it. I put my hand up to protect myself and so he hit my hand too, till the ladle broke and he stormed out of the house. After he left Mama washed the blood out of my hair. I began crying. I couldn't help it. That year had been so bad, and I told Mama that because he had beaten my hand I wouldn't be able to hold the poem and read it in school. Every year on the 6th, Santa Claus came to class with a bag of peanuts and candy for all the children. A student was chosen to stand up and recite a poem of thanks to the Queen for the gift. I wasn't the one who'd been chosen, I lied to Mama, but whether she believed me or not she pitied me. She hugged me and that was when she told me the story of my mother. I knew I was different, with a different name, Big George's name, but no one had told me the reason. Now Mama told me that something had happened back in Vrbosko. There was a carnival one year, like they always had in the spring, and Mama was sick, she couldn't go, so her husband went with her sister Magdalena. After that Mama was very angry with her sister and it almost broke up her marriage, because Magdalena was going to give birth to a child. A child that was, ultimately, me. Magdalena was so disgraced by this, so despondent and distraught, that after I was born she poisoned herself. So Mama wasn't my mother, she was my aunt, and I wasn't the real brother to Gizela and Dragitsa. Now I knew why I was different, why my name was different, and why I hadn't had a real mother. What my mother had done was wrong. I was wrong too. I was wrong for being born.

The next day, after we got our candy and peanuts from Santa Claus at school, I went down to the center of Heerlen to the big store, four stories high on the square, and bought Mama a gift. I had some money—I'd stolen it from home, if you want to call it stealing. I considered it mine because in the summer when I worked at Theo's father's farm whatever I got paid I had to hand over at home, unlike the other kids who could use theirs to buy candy and peanuts and pirate cards. I rode up and down in the elevator and looked all over the store, mingling with all these people I thought were so important, and finally I chose a glass dish with three or four crocuses in it. Obviously they couldn't survive but they looked so beautiful and Mama was very pleased when I gave it to her. She was very surprised too, because it was the first time I'd ever given anyone a present. After I gave it to her she asked me for my report card. My sisters had already brought theirs home with their glorious grades and she knew I must have gotten mine because they handed them out after Santa Claus gave us our gift from the Queen. I told her I hadn't gotten it. I think she knew I was lying, that I was hiding it, and how scared I was. It had been such a bad year at home that it had affected me mentally. There had been times that fall my father was so angry I thought he would kill me. One day he came home drunk and beat me so hard that I defecated. Then he ridiculed me for shitting my pants and dragged me into the laundry. He made me take off my pants and scrubbed me with the brush, the one I used to wash the floor. When Mama and my sisters came home he told them what I'd done and they laughed. "Josko shit his pants and I had to scrub him!" I was just completely alone, and at school I had a total blank. I couldn't think, couldn't study, couldn't do arithmetic or read or write. My report card was so bad, they were going to hold me back a grade. The shame of this was so great because I knew as soon as my father found out he would take this as a reflection on him. When he cursed me he always said I was stupid, I was lazy, that he had to feed this lazy bastard who lived in his house. "I have to feed you! Go out, clean the yard, scrub the pigeon coop!" he would shout and grab me by the hair and throw me out the back door. And he cursed my mother, the whore who seduced him, which I now understood, although I still didn't understand why he said it or why I made him so angry. But because I was being sent back a grade, what he said would be true—I was lazy and stupid, and had disgraced him. And because of the report card, the greatest joy I had at Christmas was in giving Mama the gift of crocuses.

I was constantly preoccupied with the thought of leaving home. It was my safety valve—I was going to sea, back to Yugoslavia, anywhere. When I left Vrbosko, my grandmother told me I would come back after I finished school, that it was good for me to go to Holland because I'd get a better education there. When I came back the Filipovic relatives could get me into the Yugoslavian Naval Academy. Perhaps for a while I even believed it might happen. My father did too because every day he made me study Serbo-Croatian. Despite the fact that he couldn't read or write himself, he'd make me copy passages from a Yugoslavian history book, and I had to do it every day. My sisters didn't have to do this copying, but he considered it more important for me than the homework I was given in school. The only way he could tell if it was right was to compare what I'd written to the book and if I hadn't crossed the z or t or j or put a dot on the i he'd go into a rage. It was his ego, *he* was going to be the one who taught me to write, although he couldn't read, and *he* would be the one who got me into the naval academy, though he had nothing to do with it. At the same time he resented it. He resented it if I accomplished anything on my own, so the idea that the Filipovic relatives were going to send me to the naval academy was just a dream. I didn't really believe it myself. It wasn't just because of my father. My destiny was to chop coal. We were bred for that, raised for it. The Dutch schools never let us assume we would go to high school, the *Gymnasium.* That wasn't for us. As soon as I finished 8th grade, which was compulsory, I'd go to work and bring home money. The same went for the other kids on the street too. Even *Mijnheer* Van Dijk who took an interest in individuals or the old priest who took me out for walks, they would never dare tell me I could go to high school, and rightfully so because they knew there was no way that would happen.

The year I turned twelve, the last year before the Germans invaded and the war began in Holland, was the worst year of my life. My father had had a third operation and afterwards they wouldn't even let him go down below. They offered him some kind of job that required reading and writing in Dutch—perhaps it was even a good job—but he couldn't do it. One night he sat at the table in the kitchen staring at the pages in a Dutch language book. Mama was trying to help him understand it, but it was no good, it was not going to happen. He could barely sound out a printed word in his own language, so there was no way he could read Dutch. When he went back to work they gave him a job on the wood pile. It was very humiliating, even worse than tending horses. At the time I didn't relate his illness and his own fear and humiliation to how he treated me, especially because my friends were treated differently by their fathers. Mr. Kamic sat out on his front step in the evening and he talked and joked with his sons, and Mr. Timmerman was the same with Sheng and Leo, he enjoyed them, he loved them. If they got in trouble they got a beating, that went without saying, but not with the *schlauch*, and their fathers didn't curse them and punch them in front of their friends. A girl came to the house to see me and he hit me in the face with his fist, punched me to make me cry, to humiliate me so she'd go away. If I was sick he pulled me out of bed, cursing and shouting, *I fuck your mother! Get out, get up! Get to work!* For whatever reason, his pride, his illness, his frustration, my father had become totally unreasonable, and my aunt, Mama, suffered because of him too, the way he went in a rage if he found a trace of dust anywhere, sometimes beating her too. In Frankstraat, although my sisters were pretty and did well in school, his behavior was so ugly that we were the trash on the street.

My dream of escape at the time was just fantasy, but it kept me alive, it was my motivation, to escape from my father and from the coal mines, go someplace I could live like a normal person. I told my friends I was going to get out, get the hell away from here, and I also told my first girlfriend my dream. Basha was a Polish girl who lived up the street, and one day she gave me a sheet of blue paper with what looked like a poem written on it. When I read it I recognized the lyrics to a song I'd heard on the radio. It was like a French Legionnaire's song, "He walked for many months on the wharves, and the future had no certainty...*ver van alles dat ik heb liefgehad, vergeet me niet, en denk van tijd tot tijd een ogenblik aan mij*...Far from everything I have loved, forget me not, and think of me from time to time for just a moment." When I read the lyrics, I could hear the music, and I sang it to myself on a Sunday afternoon when I was sweeping the floor and everyone was gone. The song was about someone reaching for love, and I just felt so desperate I sang it over and over at the top of my lungs. If anyone had heard me singing so loud in my off-key voice, they would have thought I was crazy. Obviously I was crazy. But there was no one to hear me. The family was gone, Sonya next door and her parents were gone, Wil and his wife on the other side were gone. The street was empty and I had been left alone to clean the house on Sunday, and it just got the best of me.

CHAPTER 5
HEERLEN, HOLLAND; SUMMER 1941

I was walking down the road on a Monday during the second summer after the Germans invaded, when a guy stopped me in the street. It was Joop, a Dutch coal miner who was eighteen or nineteen. We knew each other because he'd been the Boy Scout leader when I was a member of the group. The clubhouse was run by the priest who had asked me to join and my father never opposed the church. This group of Boy Scouts had always put on plays at the local theater and we put on a show the first year of the occupation. It was a comedy about inheritance, Joop played my father and while the family was divvying up all the "marbles" I got pissed off and said, "My father says if he doesn't get enough money, I won't get a new bicycle," which broke up the theater. That had been early in the occupation, the previous fall, and soon after that the Boy Scouts were disbanded. Later they started the Dutch Hitler Youth which of course I had nothing to do with. Neither did my friends. Scheng and Leo hated the Germans for invading their country, and Vwadzu and his brothers were Polish and Mr. Kamic had been in tears for a week when Poland fell at the start of the war. For them to join something like Hitler Youth would have meant they were collaborating, and there was no one we despised more than the Dutch *NSB*ers in their black uniforms with an orange sash. To someone like

Scheng they were traitors. I also noticed that the people who sympathized with the Germans and joined the Dutch Nazi party were the ones who were better off, and that certainly wasn't us.

When I ran into Joop this particular Monday morning, he said hello, asked me how I was doing, and then he said, "By now you've heard what I'm doing."

"No, I don't know," I said. I'd heard rumors he was involved in something. No one called it the resistance, it wasn't like the movies, no one called it anything, but I didn't know for sure. Little signs with an orange "V" would appear stuck up on fences or trees to oppose the big banners with "V" for victory the Germans had hung in the square, but there hadn't been any real sabotage yet.

"We've been meeting a couple times a week," Joop said. "We need some more people to join our organization."

"What would you want me to do?" I asked.

"We need someone to deliver messages."

"What kind of messages?"

"Envelopes that have to be delivered to someone."

"I don't have a bicycle."

"It's better if you walk," Joop said. "Because these have to be delivered at night."

"You know I can't get out at night," I said.

He knew my situation at home, my father would never let me out without a reason and I couldn't tell him I was going to deliver messages for someone. It was not just my father, it was the Germans too. There was a curfew now and no one could be out after dark unless you had a special pass to go to work. So Joop and I said goodbye and he said he'd see me again.

By the second year of the occupation things had changed. The first year wasn't so bad. The main thing was everyone became more and more conscious of food. Right after the invasion, everyone over twelve had to get an identification card from the *Deutsches Haus*. Without it you couldn't get a ration card, and that meant no food. There were shortages because a lot of the food was being shipped back to Germany. Many things had just disappeared from the stores. Like only the coal miners got chocolate. If they worked Sunday, which had become compulsory, the miner got eight candy bars and a bottle of wine, and he had to be an actual miner, one of the ones chopping coal. I used to buy candy from the miners and sell it to the upper class Dutch to make a few *guldens*, but my father never got candy because now he worked up above. The family also suggested I get milk for them, so I'd get up at two or three in the morning, grab a bucket and go out to the field across the road and chase down a cow. I'd tie her to a tree, milk her, and lose half the bucket of milk running back. An older Dutch kid named Henk and I also stole wheat. We'd go out and grab a stack tied up in the field, stick the head in a sack and chop it off. Then I'd roll the heads in my hands to separate out the chaff, roast them, and grind them in a mill. It was a lot of work but everyone was always looking out for food. The coal miners ate better than anyone—no one got pork except the miners—but during the first year of the occupation the Germans kept cutting back on the ration. So in the spring the coal miners went on strike. It wasn't for money, they wanted food, and they struck all the mines around Heerlen—Orange One, Two, Three, and Four. Immediately, the Germans arrested sixteen men, executed eight right away, and posted the names of the other eight on telephone poles with a notice saying that if the miners weren't back at work by three o'clock the next day they'd shoot the other eight men too. They said they'd keep arresting and shooting men till the miners went back to work. That was it, the strike was over. For a moment people had felt we can say something, we can be rambunctious and the Germans won't do anything. The executions ended that feeling instantly. All of a sudden it was as if there was a burden on everyone. A quietness settled over the area and people became even more anti-German. Before people had been unsympathetic and disliked them, now the feeling was hatred. A Dutchman and a German bumped into each other on the street; neither would give, so the German just shot him right there with a pistol. A priest was bicycling down the street and an *SS* man shot him. Two young guys were shot at the swimming pool because the German claimed they had spit on him. The atmosphere turned very grim. Everyone had to turn in their radios. They didn't want us listening to the BBC. The Germans in the neighborhood were allowed to keep theirs and whenever there was a *Sondermeldung*, a "Special Report", they'd throw open their windows and turn up the sound to make sure we heard it: *"Achtung! Achtung! Sondermeldung..."* The broadcasts were carefully produced with sound effects of dive bombers screeching, bombs blasting, as if the announcer was right on the spot: "I can see the factories *brennen! Bomber schlagen zu...*" It was all so theatrical, the noises so brutal, the announcer so elated—the whole broadcast was hysterical.

The German radio was also constantly berating the Jews and it was soon after the strike when they announced that all the Jews had to wear a piece of yellow cloth with *Jood* written on a black star. We had no Jews living in our neighborhood but on Saturdays a Jewish man named Kaufman came around house to house carrying two suitcases full of fabrics, dress material, and buttons, and my aunt bought whatever she needed on credit. She'd give him three or four *guldens* against what she owed and he'd mark it down in his book. He was a very kind man and my aunt always made him pancakes, and he'd sit there and eat and talk. *Mijnheer* Kaufman disappeared, most likely he was sent to a concentration camp because he had roots there, those were his people, and he never came back. He was the only Jew I saw in our neighborhood. There were synagogues in Heerlen but there were also churches so I thought no differently about them. The Dutch were very tolerant of the Jews and considered them citizens like anyone else. There may have been discrimination in Holland but as a child I didn't see it. I had seen it in Germany, before the war started, when I went over to Aachen with my aunt. The *SA* were burning a Jewish store and rounding people up. They arrested the owners and beat them on the street while people went into the store and looted, then set it on fire. They were like a pack of animals, throwing things into the burning store, and I couldn't wait to get out of there. Now it was happening here. In Heerlen I saw a truck pulled up in front of a house. There was a small crowd of people watching outside. The *SS* were searching the house and they herded the people out, beating them. The fear in the eyes of the people as they loaded them on the truck was a kind of fear I'd never seen. The people looked like absolutely, totally lost souls. While they were loading them, we heard screaming and one of the *SS* men came out of the house holding an infant by its leg. The baby was crying and he swung it, bashed it against the bumper and threw it in the back of the truck. Some of the people in the crowd started to cry. I felt sick. The fear in their eyes made me sick, like I was going to throw up.

We all knew they were taking the Jews to the concentration camp and by this time we knew that if you went to the *KZ*, the *Konzentrationslager*, you wouldn't come out. The collaborators and Germans used that threat to keep us in line. "If you don't behave yourself, you'll go to the *KZ*." The word had crept into our lives, it was always around. The Jews were going because they were Jews, but it was also for anyone they wanted to get rid of. Many of the Dutch officials had gone to the *KZ*, and Mulder's son who was vehemently anti-German and had a fight with a couple of Huns was sent to the *KZ*. They rounded up Jews for three or four weeks; after that there were no more Jews, they were gone. I don't know if the feeling toward the Germans could have been any more hateful than it already was, but now the anger was even deeper.

The anger for me had become personal too, ever since the Germans invaded Yugoslavia. It was late one afternoon in April when I heard the triumphant sound of the announcer shouting the *Sondermeldung:* "The Germans and Italians have invaded Yugoslavia..." When Yugoslavia fell just eleven days later, I had to go off by myself so no one would see me cry. What I felt was my national pride, and besides that my dream of going back to the naval academy was shattered. The idea that I could get away from my father and out of this hell hole was over; now my destiny really would be the coal mines. I detested the coal mines—the looks on the faces of the miners, the way they lived, the monotony, the drudgery—and now that Yugoslavia was defeated I felt the helplessness and the hopelessness of the way my life had been formed. Whatever way out I thought I had, the naval academy, the merchant marine, that dream was dead. Now I was going to finish school and go down to the coal mines. I'd be stuck at home and there'd be no escape from this place.

A month or so after Joop spoke to me the first time, I ran into him again. It was a Sunday morning and I was on my way to church with Vwadzu. Actually we were going to the pool hall next door to the church to play billiards and drink beer. If you could pay for it, they'd serve you, although it was so weak you could drink a barrel of it and it wouldn't do anything. Joop waved me over while Vwadzu went in the pool hall.

"We still need someone to deliver messages," Joop said. "They're watching all the people in our group."

"You know I can't get out at night," I said.

"I'll tell you what we can do. Tell your father you've made friends with a farmer who wants you to help out feeding and milking the cows in the evening. We can give you some food, some butter and wheat to take home."

"That sounds good," I said. I saw a chance to get out of the house. And I'd get my hands on some food. Just for the meal alone it was worth it, because if I did any work at a farm I knew I'd get fed.

The following Tuesday, two days later, Henk who lived across the street from me came over and said there was a guy at his house who wanted me to work at his farm. Henk said, "He's got some bread and butter for you."

I went over and met Maarten who was also about 19, the same age as Henk and Joop. They were all five or six years older than me. He gave me butter and a round loaf of farmer's bread.

"Here, take this home," Maarten said. "Tell your father I want you to work for me and I'm paying you in advance."

"Who's going to believe I got paid in bread and butter before I did any work?" I said. That was more than unusual, it was unheard of.

"Tell your parents this is my share," Henk said. "When you get paid you'll pay me back."

I said okay. They had this all figured out, and I was eager to do it.

"I'll meet you tonight," Maarten said.

When I went home and put a loaf of bread and real butter on the table my aunt and my sisters said, "God! Bread—and *butter!* Where'd you get that, Josko?" All we were getting with the ration card was margarine. When my father came home he said the same thing and I told him the story Henk gave me to use. Getting paid in bread and butter was quite a large payment but I also told him the farmer was a friend doing me a favor. "He wants me to help out tonight."

"Go!" my father said. "Go!"

No one questioned it. As soon as they saw the bread and butter on the table, they were ready to believe the story of how it got there.

In the evening I went back over to Henk's. Maarten was there and we walked to his father's farm. It was more than an hour away, seven or eight kilometers, toward Kerkrade, and on the way we talked about the neighborhood, people we knew, small talk, but nothing was said about what I was going to do. The farm was a typical European farm complex with big doors at the front, more prosperous than Theo's father's farm where I worked in the summers. While Maarten was putting his bicycle away, he said, "To make it look legitimate, you can help a little around the farm. But you don't have to work hard."

"I'm glad to work," I said. "As long as I have something to eat."

"Okay. But that's not what we need you for."

In the house I met his family, his parents and sister, and a couple of hired hands. "Josko's going to help out in the evenings after school," Maarten said.

He and I went out and pitched feed to the cows and then the *knechts* came out to milk the cows. A few minutes later, Maarten's father came out and called me over. He had an envelope, blank, no address or writing on it, and told me where to deliver it. It was a house in Beersdal. I was to ask for *Mijnheer* So-and-So, and Maarten's father described what the man looked like.

"Give it to him. Don't give it to his wife, his son, his uncle. It has to be him. If he's not there, bring the envelope back to me."

Maarten said he'd come over to Henk's with some bread for me the next day. He also gave me a coat because it was getting chilly and the trip was going to take me some time. I went north, through Heerlen, and to avoid running into any Germans I stayed in the back streets because by now it was dark, after curfew. But this meant it took considerably longer than if I'd gone on the main road. I got to Beersdal and found the house, knocked, *Mijnheer* So-and-so opened the door himself, and I gave him the envelope. By the time I got home it was one in the morning. My father woke up, came downstairs, and said, "Did you get any bread?"

"It was too late," I said. "He's going to bring some tomorrow."

"Are you sure?"

"Yeah, I'm sure he'll bring it."

"More butter too?"

"Only bread tomorrow. Maybe butter the next time."

My father started to go back to bed, then he said:

"Where'd you get the coat?"

I was still wearing the coat Maarten had given me, a brown coat. I didn't have a coat of any kind. In the winter I wore the same clothes, inside or out.

"My friend let me borrow it," I said. "I have to give it back tomorrow."

The next day Maarten came to Henk's with the loaf of bread and told me he had nothing for me to do.

"Let me work at the farm anyway, just to make it look like it's legitimate," I said. I wanted to get out of the house, and also get a meal at the farm.

"Okay," Maarten said. "You can work two days a week. Henk will let you know what days we need you."

That started the pattern that continued over the next several months. I went to the farm two days a week to work and whenever I came back I always had a loaf of bread. Or if I didn't, the next time I'd have two. Every now and then, butter. Most of the time I just worked at the farm, but every few weeks Maarten's father gave me a message to take to the man in Beersdal. Always the same, through the back streets at night, give the envelope to *Mijnheer* So-and-so, and walk home. That was it. No one at home ever suspected anything.

I never thought about why Joop had asked me to help them, but it was obvious. If there was ever anyone who wouldn't tell his family what he was doing, it was me, and Henk lived right across the street so he knew my situation. Also, perhaps because of the way I was treated at home, I was very sure of myself. I was only thirteen but big for my age, stronger than the other kids, sturdier, perhaps because I had to work all the time. Whatever the reason, I was sure nothing would happen to me. I would get away with whatever I was doing. I never thought about getting caught. I only thought about being careful. The only thing I worried about was that it would stop, that Maarten wouldn't have anything more for me to do. Because the job was important to me. It was my release from the house and my father. After a couple trips—not immediately, but after several trips delivering envelopes—I felt elated. I was doing something in the adult world. I was treated with respect. I didn't have to come home and lie and say, "I stayed in church late," or "I stayed after school and helped put away books for the teacher." I could say, "I worked, I threw hay to the cows, I swept out the barn," and when I put the bread and butter on the table that was the proof. All those years when my father cursed me and cursed my mother, the whore who seduced him, saying, "You lazy bastard, I have to support you, I have to feed you." Now I was feeding *him*. I was bringing home food for the family, and it was the kind of food no one could get because of the war. I was suddenly on a different level.

By the next spring, I'd made perhaps fifteen trips delivering messages. It was always the same, on the back streets at night, returning home at one or two in the morning with a loaf of bread and, if I didn't have bread, on the next visit to the farm I'd get two loaves. I knew my way around and I never got stopped. Then one Sunday when I went to the farm, Maarten said they had something else for me to do. There were two other men there, guys in their twenties I'd never seen before. Also a third guy who was a kind of coordinator. There were no names, Maarten just introduced me by saying, "He is the one who's going to show you the way." On a piece of paper the coordinator drew a rough map of Maarten's farm, the main road, and streets that ran off it. Then some boxes for houses. "You know the way," he said to me. "This is where you're going." He pointed to one house on his "map", and I said I knew the street, I knew where the house was.

Then he gave us directions about how to walk, very exact and precise, so we weren't all walking together. It was daytime and I left first, then they were told to follow separately. We met about three blocks away but didn't greet each other, and they were on opposite sides of the street, following me. My instructions were not to walk either too fast or too slow, and after eight more blocks they joined me and the three of us walked together. We came to the street, I showed them the particular house, then we came back to Maarten's, repeating the same pattern in reverse. About four blocks from the farm, they dropped back, and we came back one by one. I was told to come back the next night, Monday, and take them to the house. They knew the way, but they wanted me there just in case, because I knew the streets.

So the next night I came back to Maarten's farm and we did the same thing. It was dark when we got to the house again, and this time we went past it to the end of the block, crossed the street and came back on the opposite side. There was a hedge with a gate—the houses were nearly identical on both sides of the street—and one of the guys told me to open the gate and go in. They followed me in and crouched down behind the hedge. The hedge was low; you couldn't stand up behind it without being seen. Then I noticed one guy had a pistol, and when I looked at the other guy he had one too. They were both Dutch but not from the area; most likely from Brabant by the sound of their accents, and whatever they were going to do was something I hadn't been told about. We waited a long time, a good hour, hour and a half. Finally, the front door of the house across the street opened. It was blackout, no light from the windows, but there was a moon and I could see two women and two men in uniform, German officers. We could hear them laughing and saying goodnight, then the door closed and they came down toward the hedge and gate in front of the house. When they opened the gate and came out to the sidewalk, the two Dutchmen jumped up over the hedge, ran across the road and shot them. Only two shots, almost simultaneously, very fast. The Germans didn't know what hit them. The two guys picked up the bodies, packed them on their shoulder, and where the day before we had gone to the next block and turned right, now we turned toward the other side of the road where there was a gently sloping embankment. The Dutchmen threw the bodies down the embankment and they sort of rolled down toward another road. I could see what looked like a truck parked down there, but the guys didn't wait, they just dropped the bodies and kept going. We started walking back and I heard the truck take off, so I assumed someone down there had picked up the bodies. But the guys didn't say anything, and I led them back toward the farm.

At the time I just functioned, I didn't think, but the suddenness of what happened—the two guys jumping over the hedge, the shots, rolling them down the embankment—it was totally unexpected. I'd been delivering messages, now I was helping kill German officers. When I got to the farm, no one was there but Maarten. The coordinator was gone and I never saw the other two guys come back, they just evaporated. Maarten asked me if everything was okay and I said yes, but because I had been so conditioned not to say anything I didn't tell him what happened. I didn't even know if he knew what we'd done.

"I'll bring you some bread tomorrow," Maarten said.

I was so anxious, I wanted to stay overnight at the farm, but he just said goodnight. I left. I still had to walk a good hour back home. Every time a car or a vehicle passed I thought, that's it, they're going to stop me. As I walked I saw everything again in my mind, more clearly now because when it happened it was so fast and such a surprise. The Germans had been officers, I could tell by their uniforms, and there had just been one shot for each one so the Dutchmen must have known what they were doing because there was no struggle. When they rolled them down the embankment I didn't know if they were already dead but obviously it had been arranged because, again, there was no struggle, no sound, and someone was there to pick them up. I had only been told to show them the way, although there was no reason for them to tell me anything more. When it happened I felt nothing, it was like carrying the envelope. But by the time I got home, off the street, and then the next day, I could feel—it wasn't a satisfaction, because the sound of killing someone, you don't forget that—but I felt an importance, that I was doing something. On the radio, whenever I heard a *Sondermeldung* screeching about the courageous Germans fighting the bandit Tito in the mountains of Yugoslavia, I was proud. The French, the Dutch, the Belgians weren't fighting the Germans, but the Yugoslavs were. So to me what Tito and his partisans were doing was the way to oppose the Germans. Now I was part of this. I was accepted and trusted, and once again this was something completely new to me. I had fallen into a family or a community that I had never had. Although afterwards everything disintegrated, and if I was ever tempted to tell Vwadzu or Scheng what I'd done, I knew—Maarten was very clear about that—you don't say anything to anyone. And if anyone comes to get you, you're on your own.

As soon as I began working at Maarten's farm, the beatings at home became less frequent. My father still cursed me and made insinuations about my mother but the physical abuse was evening out. I was bringing home food, and a loaf of bread was more valuable than the wages I could have earned in a week at the coal mines, so my relations with my father were changing. But in the spring when I finished school I was confronted with the problem of what I was going to do. I was fourteen now and some of my friends had already signed up to work in the coal mines. Others were going to trade schools and now that I had no chance to go to the naval academy in Yugoslavia or join the Dutch merchant marine, I was hoping to go to the *Ambachtschool.* One evening after school ended my father was sitting on his couch after supper and he said, "Josko, come here—I want to talk to you."

The couch he always sat on was on one side of the table where we ate and I sat down in a chair across from him. He had never spoken to me seriously before. When he spoke to me it was a curse or a command, so the way he was treating me now was something he had never done before.

"When I was a young man your age," he said, "I went out into the world and worked. Ever since then, I've made my own way. Now I want you to go out and learn what it's like yourself. I want you to learn a trade and pay your own way. You're too old for me to support you any more."

It made me so angry to hear this—as if I wanted him to support me. I wanted to get away from him. But when he said he wanted me to learn a trade, I thought he was going to let me go to the *Ambachtschool.* If I did learn a trade and got a job, all my pay would go to him till I got married, just the same as if I went to work in the coal mine.

"So," he said, "today I went to the *Deutsches Haus* and told them I have a son who's just finished school. I gave them your name and told them you want to work."

When I heard that, I wanted to kill him. He had *volunteered* me to work for the Germans!

"They said they would send you a notice to tell you when to report," he said.

He was turning me into a collaborator! If you were unlucky enough to get picked up and shoved in a truck and shipped to a labor camp, that was different. But to volunteer to work for the Germans was such a cowardly thing to do. You were saying, "You are very nice people to give me a job and I'm glad to work for you." I never hid my anti-German feelings at home, especially with my sisters, so he knew what it meant to betray me to the Germans. It meant he was trying to destroy me.

At home I was the only one who was so anti-German. It was not only the shame and frustration I had felt when Yugoslavia was defeated, it was also the brutality of the Germans, the arrogance and belligerence, the screeching of Hitler and Goebbels on the radio, always shouting that they were *Das Herrenvolk,* the master race. I reacted very strongly to all that, perhaps because of the way I was treated at home myself. My sisters were different. They were pretty young girls, they wanted to dance, go to movies, meet young men, and they didn't care. Gizela was eighteen and Dragitsa sixteen and although there was a war going on, if they saw a good-looking soldier it didn't matter to them if he was German. Or Dutch or French or Belgian. A soldier was a soldier. During the first year of the occupation, they came home excited one day because in the railroad station in Maastricht they had seen a German war hero. Dragitsa had to tease me about it because I was very protective toward them, even though they were older, and she knew she could get me going.

"Guess who we saw in Maastricht, Josko."

I thought maybe they'd run into someone from Frankstraat like Leo or Henk.

"The *Dolle*," I said. The *Dolle* was the crazy Hungarian kid who lived in Frankstraat and ran up and down the street yelling, 'Hi-yo, Silver!' like the Lone Ranger.

"The *Dolle*!" they laughed. "This guy was handsome. He was wearing a uniform."

I always told them not to talk to any German soldiers. They knew how angry it made me.

"He had so many medals there wasn't room for them!"

"Some old man," I said. "Some old *Mof* you were staring at."

"Old?" they laughed. "He was young," Dragitsa said. "A *Ritterkreuzträger!"*

The *Ritterkreuz* was the Iron Cross and the Germans only gave it to heroes in battle. The guy was a pilot—we'd seen him in a newsreel at the Royal Theater on the square in Heerlen.

"He talked to us, Josko! He asked where we lived and if he could visit sometime."

The pilot was a real hero but to me he was the enemy.

"You talked to that Hun?"

"Oh, yeah—he wants to see us again."

"If I was there you wouldn't talk to any *Moffen*!"

"What would you do about it?" she taunted me.

"I'm warning you," I threatened her. "Don't let me see you talking to any *Moffen!"*

"There's nothing you can do about it!"

I gave her a push and she fell back against the kitchen table.

Then I heard someone shouting, cursing me—it was my father.

I was so angry at Dragitsa I hadn't heard him come home. He came across the kitchen and grabbed me by the neck.

"She was talking to a soldier, a *Mof*!" I shouted. I didn't mean to tell on her but I was so angry it just came out.

"Is that true?" He looked at Dragitsa.

"Josko always tells lies," she said.

"Were you talking to a soldier?" he shouted.

He let go of me and pounced on her. He never hit the girls, never touched them, but they were older now and he was always suspicious when either of them had been out—where had they been, what were they doing, who were they with? He went berserk. He pounced on Dragitsa like a vampire, his hands were claws. He grabbed her by her hair, her long black hair, and yanked her back and forth while he slapped her, snarling and cursing like an animal while she screamed.

I stood there and watched. It was always my sisters who got me beaten by telling my father I'd done something to them. They threatened me with it when they wanted to punish me, but now Dragitsa was the one being beaten and I felt ashamed I'd told on her. I hated him for hurting her, but at the same time, because of the way I'd been taught, for a moment I even thought he was entitled to do that because he was the father and she was the daughter. Then I felt my anger towards him and wanted to make him stop. But I was so young and afraid I did nothing. I stood there and watched.

The next time I went to Maarten's farm, I decided I would tell Maarten my father had given my name to the *Deutsches Haus* to see if he had any advice—maybe they had somewhere I could go, maybe I could work for them in some other part of Holland. But when I got to the farm, before I had a chance to say anything, Maarten asked me a question. "Josko, you speak German, don't you?"

"Yes," I said. I spoke it pretty well, with the local accent, and we'd also learned High German in school.

"How would you like to wear the Hitler Youth uniform?" he asked.

I thought he was kidding.

"What for?"

"We want you to direct traffic," he said.

"Direct traffic? How would I do that."

"It's easy," he said. "Come on."

We left the farm and walked towards Heerlen. After a couple kilometers we came to where there was a bridge, a hundred meters or so up ahead, crossing above the road we were on. It was on the main line of the railroad that ran from Maastricht near the Belgian border, through Holland and continued across the border to Aachen.

"Tomorrow we're going to put you under the bridge in a Hitler Youth uniform, and when you see any traffic coming all you've got to do is yell at them to keep moving. Just wave them on so they don't stop."

"What're you going to do?"

"You'll find out tomorrow," Maarten said.

We went back to the farm, I did some chores, and Maarten told me to be there the next morning at seven.

The next day was Saturday and I was there early. When I knocked on the door to the farmhouse, Maarten opened it—wearing a German army uniform. I went in and the whole room was filled with Huns! There were eight young guys, all dressed in proper *Wehrmacht* uniforms, the regular army. The only one I'd seen before was the guy who'd given me directions to guide the two men from Brabant. He was the only one not dressed in a uniform and once again, he was in charge, giving instructions. Maarten gave me the Hitler Youth uniform—short black corduroy pants, brown shirt, swastika, a belt across my shoulders and a sidearm, which was a knife with the swastika emblem on the sheath. I also got a hat—not with the orange slash like the Dutch Hitler Youth, this was the German one, the real thing. Everything fit perfectly, like it was made for me.

The coordinator came over and said to me, "You're going to walk in front of the 'soldiers' and lead them to the bridge. Just keep walking and don't stop. If anyone stops you—an *NSB*er or a *Mof*—tell them you have to take the soldiers to the barracks."

"The barracks?" I said. "There aren't any German barracks near the bridge."

"Just tell anyone who stops you're taking them to the barracks. When you get to the bridge, wave on the traffic so no one stops, and if a German car or military vehicle drives past, be sure you salute. Give them the *'Heil Hitler'*."

We all went outside and the scene would have looked funny to me if I hadn't been so apprehensive—all these "Huns" running around in *Wehrmacht* uniforms. They went in the barn and came out with four large German ammunition boxes. I couldn't believe they had all this equipment because it all looked so real—it *was* real. One man took either end of each box and we all went out the front gate to the road. They formed a line with me at the head and I started leading them toward Heerlen. It was about eight o'clock by now, a bright sunny day, and as we walked all I thought was I hope we don't see anything on the road—not a car, a bicycle, not anyone—because if anyone stopped they'd talk to me first. I was also the one who had to salute because I wasn't carrying anything. There was a lot of traffic, people bicycling back and forth in both directions, as well as pedestrians. Most of the Dutch had such scorn for the Hitler Youth uniform that the people passing by wouldn't even look at me—fortunately, because I saw several who might have recognized me if they'd taken even a quick look.

When we got to the bridge the "Germans" all scrambled up the embankment with the ammunition boxes. They opened them up, got out their tools, and a few minutes later they were climbing all over the bridge like monkeys. They were hammering and banging on it like they were fixing something, all the time yelling at each other in German: *"Mach schnell! Schnell!"* They did a good job of making it legitimate. I stood on the road down below, waving the traffic on and praying no one stopped. I didn't know what the hell they were doing—although by now I had a suspicion—and I kept glancing up at them to see how much longer they were going to be working because almost as soon as we got there I had to go to the bathroom. The urge was terrible, I was dying—out of fear. To keep from pissing my pants I waved my arms at the traffic like a windmill, but it really wasn't necessary because no one was going to stop to watch a bunch of Huns working on a bridge. I'd been standing there about half an hour when suddenly I saw the one thing coming toward me I didn't want to see. It was a German army truck full of soldiers—

real ones. I shouted up at the guys on the bridge, "Here they come!" but they didn't hear me. I waved my arms twice as fast as before and when the truck got close and I saw the driver and *korporal* riding up front, I shot my arm in the air and gave them the biggest *Heil Hitler* they'd ever seen! They hardly noticed me and just gave a bored salute back, acknowledging this little shit kid, and the truck barreled on down the road under the bridge.

We had gotten there shortly before nine and finally, about eleven-thirty, when I thought my bladder must have exploded, I saw some of the guys coming down. If anyone had been watching, they would have thought those were the hardest working soldiers they'd ever seen because no one had even taken a break to smoke. Maarten came down the embankment and said, "Okay, take us back."

I marched them home, carrying the ammunition boxes. Once again nobody paid any attention to us. If anyone had seen us going into the farm they might have thought Maarten's family had become collaborators with all these Huns hanging around the farm. The men took off their uniforms and just like before with the ones from Brabant, they disappeared. I didn't even see them leave. By the time I took off my uniform and gave it back to Maarten it was close to one o'clock. I asked him if he had any butter. I figured a job like this was worth butter.

"I can give you some, but I'll bring it down to Henk's later," he said. "I don't want you walking on the street in daytime with it." Everyone knew Maarten lived on a farm so he had butter, but what would I be doing with it?

"What about the bridge?" I asked him.

"We're going to blow it up."

That's what I'd guessed, although I really didn't believe it.

"Three o'clock," Maarten said. He told me someone would go back and warn the people who lived in the houses by the end of the bridge just before it went off, because it was going to be a big blast.

"We're going to lift it off," he said.

When I left the farm I was in no hurry to get home, so I went to see a friend in Huiskens, another neighborhood full of coal miners. By the time I walked back up the road toward Frankstraat

I knew it was close to three o'clock, and I kept looking in the direction of the bridge. After the experience with the two men from Brabant, I was sure they knew how to blow up a bridge, but because of my fear of getting caught doing something so big I almost hoped it wouldn't happen. I was at the bottom of Frankstraat where it joined the main road when all of a sudden I saw bright flashes of light and immediately beams went flying and all kinds of junk was up in the air. It was about three kilometers away, but Holland is so flat I could see the explosion, and on the way down the sound reached me: BOOOOM!! As far away as it was, the explosion rattled doors and windows on Frankstraat and I thought, that's it, they're going to be here to arrest me in ten minutes—there was just no way we'd get away with it. People were coming out to see what the hell happened. They were looking around, shouting and asking each other questions.

"What was it?"

"I heard an explosion!"

"What happened?"

There was a huge cloud of smoke and dust rising in the air where the bridge had been.

Because of my anxiety I had to be alone. I was almost in front of Vwadzu's house by now and was afraid he or one of his brothers would come out and say something to me, so I crossed the road and went into the wheat field to be by myself. After a couple hundred meters I came to the railroad tracks that ran into Heerlen—the same ones that went over the bridge—and started walking along the ties. I began running I was so excited. Now I didn't have to hide my anxiety, or my elation, and I felt like shouting my own *Sondermeldung: Achtung! Achtung! Saboteurs and enemies of the Fatherland blew up the bridge on the main line from Maastricht to Aachen!"* When Maarten said they were going to lift it off, that's just what he meant. Those "Huns" really knew what they were doing—although I never believed it till I saw the flashes and beams flying and heard the boom.

I waited till the next day to go back and look at the bridge. There was no bridge. Just rubble all over the place and the rails were twisted like pretzels. The Germans, real ones, were swarming around, working like hell to fill in the road down below. They brought in railroad cars from the coal mines filled with waste rock and dumped them into the road, closing it off. Then they laid new tracks over the fill. They must have figured it wasn't worth rebuilding the bridge because whoever did it would just come back and blow it up again. I also looked at the houses. The one nearest the end of the bridge had been blown up, the roof was gone, the sides caved in, and the next two houses had been hit pretty badly too. A couple days later, Maarten and I were walking together and we passed the bridge from a short distance. We saw the twisted metal beams and rubble lying in the field around where the bridge used to be, and laughed. But it wasn't funny for the people whose houses we'd destroyed and I felt badly for them. No one was hurt because a few minutes before three o'clock someone went to their houses and told them, "Go away, get the hell out—there's going to be an explosion!" and they did. But they were mad. Even if they hated the Germans, those were their homes and the Germans certainly weren't going to rebuild them. While we were walking back to the farm I asked Maarten why we hadn't blown up the bridge right away, after we got back. "Why'd you wait till the afternoon?"

"Some of the guys had to get away from here," he said. "They weren't just casual coal miners. They've got reputations."

They had obviously been trained for a purpose and might have been known to the Germans as saboteurs. Right after the explosion the Germans had set up roadblocks and sealed off Heerlen in and out, so if any of them had been seen they might have been recognized and arrested. But they had a head start and by the time the bridge went up they were gone, long gone.

A week or so after we blew up the bridge, I got the notice. My aunt had already opened it and she handed me the envelope. I had never gotten a letter before and it was intimidating to see my name typed out and to read the official document with the sign of the *Deutsches Haus*: "Josef Filipowic is hereby required to report in four days..." It had been several weeks since my father told me he gave them my name and I hadn't heard anything. I thought maybe something had happened—the *Moffen* had forgotten or mislaid my name. A stupid hope. As intimidating as it was, getting the notice also made me very angry at my father again, and at the Germans. I still hadn't said anything to Maarten about it so when I went to the farm in the evening I told him I'd gotten a notice to report to the *Deutsches Haus*.

Maarten didn't look that surprised; he just looked thoughtful and nodded.

"My father gave them my name," I said, so he'd know it had nothing to do with anything else, that they didn't suspect me of anything. Of course if they did, they would have just come and picked me up.

"They told me to report in four days," I said.

Maarten cursed the *Moffen* and gave me his sympathy, but he didn't say anything else. He didn't offer to put me in hiding or suggest any place I could go, he just said it was too bad and wished me luck. "*Geluk*, Josko."

When I left the farm, I felt almost as if Maarten had betrayed me too. Nothing like what I felt about my father giving my name to *Deutsches Haus*, that was different. But I knew if I'd been caught working for Maarten I'd be dead. Now I needed something and all they could say was "good luck." I was alone. Completely alone. Walking along the road I heard the song in my head, the lyrics written on the piece of blue paper Basha had given me: *"Ver van alles dat ik heb liefgehad, vergeet me niet, en denk van tijd tot tijd een ogenblik aan mij..."*

The next day I went to the school to see the priest, Father Himmelreich, because I knew his attitude. Although he had a German name, he was Dutch and had always taught us to oppose anything the Germans did. He told us to pray that our rosaries would become bullets shooting down their planes. I told him I'd gotten the notice and he said, "Don't go, Josko. Don't cooperate with them. Don't do anything to cooperate with them, whatever it is."

That was what I wanted to hear, but he didn't have any other advice about what I should do.

I spent the day walking around, avoiding the house, and in the afternoon I was in the center of Heerlen. Approaching the square I was very careful because the Germans sometimes picked up young guys on the street for labor camps. I was still a little young for that but they weren't all that discriminating when they grabbed people. A few soldiers were around, just guarding the square and showing their presence, but things were calm and peaceful. It was a bright September afternoon and the huge banners with the giant orange "V" the Germans had put up were hung around the square. I started to cross it. There was a large open area in the middle and on one side was the streetcar depot where the trolleys came in from the various lines, and next to the depot was the railroad station. As I crossed the square I heard a train coming in, a German army troop train I assumed because a moment later a lot of soldiers came walking out of the station, heading into the square. They were *Wehrmacht*, regular army, not the *SS*, and they weren't picking up people or paying any attention to us. There was no fighting anywhere near and they looked like they were on leave so no one expected any trouble. I was crossing the square in the same direction, sort of floating along with the soldiers but keeping my distance. I wanted to see what they'd do because on the other side of the square there was a line of people waiting to see a movie at the Royal Theater. One of the people in line looked like Gizela. I walked a little faster to stay ahead of the soldiers as they came to the street that ran between the other side of the square and the sidewalk. There was a railing along the side of the square and I stopped when I got to it. Now I could see Gizela, it was her, standing in line with her friend Katrina. Some of the soldiers were already crossing the street and some of the more rowdy ones were swaggering and shoving each other and shouting to the girls in front of the theater. I saw Gizela looking in my direction and she reacted as if she'd seen me too. Then one of the soldiers walked up to where she was standing and blocked my view of her. Instantly, I felt my anger. I didn't want any *Mof* talking to my sister, so I pulled myself up on the railing so I could see. I wanted to run over and tell the soldier to get the hell away from her, but I was just a kid and by now the soldiers were flooding all over the sidewalk, talking, laughing, joking, and making comments to the girls in line. I moved down the railing so I could see Gizela because there was a commotion, something was happening. Another soldier had the first one by the arm, pulling him away. The *Mof* shook him off and now they were shouting. I could hear Gizela's voice shouting too. I started to jump over the railing, but I didn't. I stayed where I was. Then the soldier swung his arm and slapped her. Gizela hit him back. He punched her. He punched her again and again. She fell down and the line of people got out of the way. The soldier was kicking her and people were screaming. The other soldiers finally pulled him away and now all the soldiers were leaving the sidewalk, disappearing.

When I heard the siren, I jumped over the railing and ran across the street. She was curled up on the sidewalk and I threw myself on her, hugging her, kissing her.

They picked her up and put her in the ambulance and drove her away.

I walked for hours. When I came home my aunt was sitting in the kitchen with Dragitsa, crying. I didn't see my father. Dragitsa said, "Gizela's in the hospital—she was beaten by a soldier." I wanted to ask how bad it was, if she would live, but I just went to my bed. When I saw the soldier hitting her I didn't want to admit it was Gizela. I knew it was my fault. Gizela knew I was watching when the soldier went up to her. She didn't care what he was but she knew what I would think if I saw her talking to a Hun. She was the one who stood up to him. I didn't. I should have protected her, and I watched while he beat her. We all watched and did nothing. We were all cowards. A hundred of us had seen it and done nothing.

On the morning of the fourth day, when I was supposed to report to the *Deutsches Haus*, my father didn't say anything as he left the house but he stared at me and I knew what it meant: "I won't see you here when I come home." I stayed away from the house all day. I was afraid that when I didn't show up they might send someone to pick me up. I didn't come home till after my father had come home from work and when I walked in he glared at me and said, "What the hell are you doing here?"

I stared back at him and said, "I'm not going." I was daring him to do something about it. I was so angry at him I wanted him to try, because if he insisted I go—if he tried to make me—there was going to be a confrontation between us. He couldn't say, "Josko, get the *schlauch*," because he knew I'd refuse.

"You'll get us all in trouble," he said.

"I don't think so," I told him. "If they come after me they won't bother anyone else."

He didn't say any more. He didn't push it. It was the first time in my life I'd ever defied him. He could sense something different in my attitude.

The next day I left the house again. I was still worried the *Deutsches Haus* might send someone for me but I also had to leave because they were bringing Gizela home from the hospital and I was afraid to see her. On my way down the street I bumped into Kurt, a German kid a few years older than me whose family had lived in Frankstraat since long before the war. Kurt was sweet on Dragitsa and she'd told him about the notice I'd gotten, so when he saw me he said, "Josko—how come you're still here?"

"I'm not going."

"If you know what's good for you, you better go." Kurt wasn't that sympathetic to the Nazis so he wasn't making a threat, he was just giving me good advice.

"They can go to the balls," I said.

That night my father said nothing, he ignored me as if I wasn't there. Gizela was upstairs in the bedroom. I heard my aunt talking so I knew her condition. She had a fractured skull, a broken spine. She couldn't walk or talk. When I went upstairs I went by the door without looking in.

The next day I was helping my aunt with the washing machine, cranking it for her, when a second notice came from the *Deutsches Haus*. It said I had to report the next morning. I went to see Father Himmelreich again. He said the same thing he said the first time: "Don't go, Josko."

"I'm afraid this time they'll come and get me."

"Maybe you can hide out somewhere," he said. "The one thing you can't do is go to work for them."

That was the way I felt too but where could I go? I remembered Henk saying guys were hiding out in the water tower in the orchard. He said people sometimes took food out to them at night, so I decided that's what I'd do. In the morning I'd go to the water tower. I went to see Basha to say goodbye. Mrs. Wajda answered the door and spoke to me in Polish. "Do you have to go away, Josko? I heard the Germans called you."

"Yeah, I got a notice," I said.

Basha and I went out for a walk, away from the direction of Heerlen, toward Hoensbroek where there wasn't any traffic. In the pasture where I'd chased down the cows to get milk there were still some trenches the Germans had dug during the invasion more than two years ago. Now they were just holes in the ground with weeds growing around them. We walked past them and I told Basha, "I'm going away tomorrow."

"To the *Deutsches Haus*?"

"No. I don't know where, I'm just going, away." From where we were walking I could just see the water tower through the apple trees in the orchard.

"I won't be back for five years," I told Basha. I was going to be fifteen in just a couple weeks. "I'm not coming back till I can walk down Frankstraat like a man."

My father had already come home from work and seen the second notice from the *Deutsches Haus*. It was lying on the table in the kitchen and he knew what it was when he saw the envelope. But he didn't confront me about it. I could have told him I was going to leave the next day and he would assume I meant I was going to report, but I didn't. I wasn't going to give him that satisfaction. Instead of saying anything to me, he started asking his wife why Dragitsa wasn't home.

"She's gone to get medicine for Gizela," she told him.

He wanted to know when she left, how long she'd been gone. She had to go to Heerlerheide to the pharmacy, so it was more than an hour just to get there and back. But he was always suspicious when she was out, and he paced around, cursing, talking about how much it would cost, how it wouldn't do any good anyway.

When Dragitsa came home she had a small bottle wrapped in paper but he had to challenge her right away.

"Where the hell have you been?"

She gave the excuse that she had to wait, there were other people ahead of her.

Perhaps it was the tension of Gizela lying upstairs or that he was unable to take his frustration and anger out on me the way he'd done for so many years, but he wouldn't let up on Dragitsa. He told her she was lying, that she had stopped to talk to some man, he called her a slut just like her sister, and when Dragitsa started shouting back at him, he grabbed her. She dropped the bottle and it hit her foot before it hit the floor so it didn't break, but she screamed. He slapped her. His wife shouted at him to stop but he was already slapping and cursing her. He snatched her hair and started yanking her back and forth, hitting her.

I attacked him.

I grabbed him while he was hitting her and threw him in the corner of the kitchen. I pushed him down in the corner and hit him. He couldn't get up, I had him trapped in the corner. I was an animal. I had lived every day with the thought of killing him and I never thought of it as a crime, only as getting even. Now I was doing it. Dragitsa was beside me hitting him too. All the times she and Gizela had told their father on me to get me beaten—now when I was finally beating him she was helping me.

My aunt was screaming.

"Don't kill him, Josko! Don't kill him, please!"

She had seen how brutally he treated me all the years and now she was pleading with me not to kill him. Her screaming only made me hit him harder. He was still bigger and stronger than me but my anger—my anger at him, at what he had done to my life, and then betraying me to the Germans—it made me so strong he couldn't resist me. I had him trapped in the corner. I heard him gasping and trying to breathe. He couldn't get up; I was killing him.

Someone grabbed me from behind and dragged me backwards. I struggled against him but he had my arms twisted around behind my back. It was Wil, from next door. Then someone grabbed my legs—Leo—and they carried me out of the house, through the yard and into the street.

Outside, Wil and Leo held on to me. I was trying to get back in the house. Wil had heard the screaming and cursing and beating for years. He knew what I'd do if I had the chance. When they finally let me go we stood in the street in front of #22. A few people walked by, on their way home. After a while Leo said goodnight and went down the street to his house at the bottom of Frankstraat. Wil asked me if I wanted to come into his house, to stay there. I said no. Finally he went in to eat supper. I stood in the street looking at the house, thinking tomorrow I'll be gone—whatever happens, I'll be gone from here forever.

When I went back inside my father was sitting on his couch in the kitchen. His shirt was bloody and his face was cut and bruised. He got up, came over and hit me in the mouth with his fist. It was hard, I tasted the blood, but I didn't move. I didn't hit him back. I just stood there staring at him.

"You can't do that," I said.

He stared back at me but I wasn't afraid of him now. I knew he wouldn't hit me again because he knew what I was telling him: 'You can't beat me anymore. You can't touch me because now I can beat you.'

He went back to his couch. As I watched him sitting there he looked smaller and I suddenly realized I had beaten him. But I wasn't the victor. I felt sick. Because beating him was never what I wanted to do.

When I finally fell asleep that night I woke up again almost instantly. I heard sounds of shouting, yelling, boots crunching on the gravel in the street. The voices were German. A truck engine was whining, crawling up the street. I jumped out of bed, grabbed my clothes, and bumped into my father coming out of his room. We ran downstairs, through the laundry and out the back door to the air raid shelter I'd dug in the yard. We climbed down in the hole and slid the boards back over the top to cover it. A couple minutes later we heard them pounding on the front door of our house while at the same time footsteps came down the side of the house toward us. They stopped. Then they pounded on the back door. The German said, "Where are the men of the house?"

"My husband and son are at work in the coal mines tonight," my aunt told them.

I listened to hear if the German was going to search the house but he just waited a moment, then walked back to the street. We could hear more voices, more shouting, the truck moving, but they didn't come closer again. We waited. My legs ached from not moving because I had put straw in the bottom of the hole and if you moved at all it made a sound like a machine gun. It was total darkness and quiet now but we didn't move. I could hear my father's breathing. It sounded like he was making an effort, working or walking up a hill. Ever since he got sick he didn't breathe right, even after the operations. Now it was all I could hear, the lungs straining. It had been quiet outside for a long time. We climbed out, put the boards back, and ran in the house. My aunt was waiting in the kitchen.

"I told them you were at work," she said.

My father didn't say anything.

"I'm going to the *Deutsches Haus* in the morning," I said.

"Good," she said. "You better."

I woke up before dawn. As soon as I saw the first light I got up, put on my shirt and the one-piece overalls I had worn since I outgrew my short pants. The overalls were too small because I was growing out of them too. I put on the shoes my father had made for me out of a scrap of conveyor belt. My ankles were raw where the rubber sides cut into my skin, but that was it, those were my shoes.

I grabbed a hunk of bread and left the house. I had 35 cents in my pocket and no idea what would happen after I went to the water tower, but I had no apprehension whatever. Walking down Frankstraat I thought, I'm free, I'm away from my father. I would never see him again—and as bad as it was when I ended up in the camp, as miserable as that existence was and in spite of the brutality of the Germans, for me my life began the day I left home.

Chapter 6
Heerlen, Holland; September 27, 1944

There are some dates in your life you always remember. May 10, 1940 was the day the Germans invaded and the war started in Holland. September 27, 1944 was the day we were liberated. Sometime between the second and the fifth of September, Leo and I had escaped from the labor camp, so by the twenty-seventh we'd been hiding in the water tower for more than three weeks.

It was a Sunday and we heard guns, heavy firing maybe thirty kilometers away. From the rim of the tower we couldn't see whether the Germans were still at Theo's farm but from the sound of the howitzers blasting we knew something was happening. Everyone wanted to be first to make contact with the Americans, so even though we didn't know if the Germans had pulled out or if they were going to make a stand, as soon as the first guy climbed down from the tower, we all did. "Let's go to Huiskens," I said to Leo. It was Sunday so a lot of people were going to be out and there was less chance we'd be recognized there instead of Frankstraat, in case it took the *Amis* a while to get here. Huiskens was another, older development mostly of coal miners, and when we got there people were already out in the street, mingling and talking, wondering what was happening. They were still cautious and watchful because nothing had happened yet, but already the atmosphere was different. No

*NSB*ers were out in the black uniforms and shotguns breaking up groups. The air was alive, everyone was just waiting to break out. Leo and I walked up to some people gathered in front of a house and asked if they'd heard anything.

"I think they're in Valkenburg," a man said.

"When did you hear that?"

"This morning. If all goes well, they should be in Heerlen any time."

It was about five in the afternoon by now and we stood there talking, everyone so eager but not sure what to do, when a man rode by on a bicycle and shouted, "I just saw the *Amis*—they're coming into Heerlen!"

That was it. The word spread like wildfire and everyone headed for Heerlen. It was about two kilometers away and suddenly the street was full of people, hundreds of us, walking, running, laughing, talking. It felt like we were moving on air, you weren't even aware of your legs. All this time you'd been scared, you had to look out, watch what you said, who you talked to, where you went, and now we were all out in the open, rushing to greet the *Amis*.

They were already in the square. Tanks had taken up a position and there were vehicles everywhere—jeeps, trucks, half-tracks full of infantry soldiers—and everyone was trying to get to them, touching them, hugging them, climbing all over the tanks, waving and cheering. People had bottles of wine and Bols gin they'd saved for almost five years and now they'd broken them out and were handing them up to the soldiers. Everyone was out, mothers with babies and old people who hadn't left the house in years, wearing their Sunday clothes to show off the for *Amis* and crying they were so happy. There had never been so many people in the square in the history of Heerlen. Someone had already ripped down the huge orange Nazi banners and people were singing the Dutch national anthem, waving the Dutch flag. One old man had wrapped himself up in a huge flag and he was dancing around while he sang. Although this was the front line—if there was any German resistance the Americans would have to face it—the soldiers responded to all the craziness. They handed out chewing gum, chocolate, American cigarettes, all these things no one had seen during the war and they were just tossing them out to people like it was nothing. Some of the soldiers were quiet and reserved, watchful, as if they had seen a lot and still had a lot more to go, but they were enjoying it too. Because people just couldn't get close enough to them. To us the *Amis* looked like giants. They were such huge men that the rifles in their hands looked like toys, and when you saw them you knew there was no way the Germans were going to be coming back. It was not only their size, it was they way they behaved. When you saw an American soldier walk and a German walk, it was like two different people. The Americans were loose, their boots didn't have hobnails so they didn't make any noise, and unlike the Germans who always looked erect and coiled, ready to shoot, the *Amis* were so casual I couldn't believe it. Slouched in the back of the half-tracks as the column moved through the square, they didn't look like soldiers at all. They were all wearing windbreakers so you couldn't see any insignia and they all looked the same, whether soldiers or officers, but with the Germans you were constantly aware of the amount of silver trim on the collar and the rank of the officer. The *Amis* were some kind of men we had never seen before. To us they were the elite, the ones who were winning the war. It wasn't the British, the French, the Canadians. It was the Americans. They were our saviors. Everyone had been waiting so long, so many years—now they were here and the feeling was even more than we could have imagined. The word "liberation" says it all. We were free.

As Leo and I moved through the crowd, I kept looking out for people I knew, friends, to see who had survived. I looked for people in trouble with the authorities, people I would have ignored before because of things they, or I, had been involved in. It seemed so strange that suddenly there were no Germans—or *NSB*ers—anywhere. The square and the train station had just emptied out. We were so close to the border that maybe the Germans had decided to pull back and not make a stand here. Whatever the reason, they weren't fighting in Heerlen.

As festive and crazy as everything was, something else was happening too. The owner of the big hotel in Heerlen was a collaborator and someone had already chased his wife and daughter down the street, grabbed them and cut off their hair, shaved their heads.

Some of the *Amis* had driven through the square and were heading down the main road toward Heerlerheide and Frankstraat, and Leo and I took off after them. The first person I saw when we turned up Frankstraat was Broertje, a red-haired Dutch kid who'd worked in a volunteer labor unit building pillboxes and tank traps for the German army. I said, "Broertje, come here."

He took a few steps toward me and stopped, so scared he couldn't walk.

"Where's your uniform?" I said.

"Josko, I made a mistake." His voice was shaking.

"I know you did."

I started toward him when a black car came racing down the road, turned the corner, and slammed on the brakes. It stopped and a guy jumped out holding a gun. He was wearing a helmet and had an orange band on his sleeve. The driver and a third guy got out too. They all had guns. I whistled. The first guy whistled back, even before he saw who was whistling, and looked around.

"Godverdomme!" Maarten shouted. "Josko—you're still alive!"

"Yeah—so are you!"

Maarten walked up with a big grin on his face.

"Where the hell were you?" he said.

"Germany. In a camp."

"You got out?"

I told him Leo and I had escaped.

Maarten stood there grinning and shaking his head.

"Come on," he said. "We're going to pick up some *NSB*ers."

He opened the trunk of the car and took out two more guns. He gave me one and said to Leo, "You want one too?"

"Yeah," Leo said. "But I don't know how to shoot it."

"Don't worry," Maarten said. "There're no bullets in it anyway."

"Look who I ran into," I pointed at Broertje who was just standing there, too scared to move.

"The bastard," Maarten said. He told Leo to guard him and we'd be back with some more.

He and I and the other two guys started up Frankstraat. I knew where the collaborators on the street lived but this wasn't a casual round up. Maarten had a list of names of the ones he was looking for. The first house we went to was a Dutchman's. "Go around the back in case he runs out," Maarten told me. I ran along the hedge and waited. Maarten knocked on the front door and a couple minutes later he whistled again and I ran back to the street. They had the guy, he'd been sitting there, just waiting for what he knew was coming. Where else could he go? The Germans were gone, he had no one to protect him. Now he was like we'd been before—trapped—and he was terrified. With good reason. He knew what had been done to us and expected we'd give him the same. One of the other guys shoved the *NSB*er down the street.

We collected two more collaborators and it was the same thing, they were just sitting there waiting. There was no place in all of Holland for them to go. A couple more guys had joined us too, one of them I recognized from the water tower, and when we got to the top of Frankstraat he pointed to the second house in a block of four. "That one." I went around to the back again and waited, pointing the gun at the back door. They knocked at the front, no answer, knocked again, and suddenly the back door opened and a man came running out. He turned toward the hedge, saw me, and stopped. It was the same old Dutchman who'd been there when the German soldiers caught me and Leo sneaking back to his house for food two weeks ago. If it hadn't been for the soldiers, this old Nazi would have turned me and Leo in to the *Deutsches Haus* to be shot. Now I had him. I saw his fear. He knew I could do it. I wanted to kill him, I wanted to kill him more than anything, and if I did no one would question it for a minute. Neither Maarten nor the others, no one would say a thing.

I whistled.

Maarten whistled back.

"I've got him," I shouted.

Maarten came through the yard and we marched the *NSB*er down the street. As we walked I realized if I'd shot him I would have let him off the hook. The worst punishment he could suffer, worse than death, was to live, because starting right now his life was over.

After we rounded up everyone on Maarten's list, we marched them out to a trade school on the outskirts of Heerlen where they had set up a makeshift jail. The collaborators all had the same look on their faces and although some of them got pretty rough treatment, we didn't shoot anyone. As much as we hated them for what they had done to us, Maarten and the rest of us weren't killers. We weren't the conquerors, we were the liberators.

That night was the greatest party in the history of Frankstraat. All up and down the street people were celebrating and everyone wanted the same thing—to claim an American soldier and get him into their house. Timmerman's and the house on the corner across the street were the center of the partying because the *Amis* had parked their trucks and half-tracks along the road at the bottom of Frankstraat and set up camp in the field, so all they had to do was cross the road and knock on the door. It was still blackout—now it was against the German planes instead of the Allies—but no one paid much attention to the rules because we were all so crazy with excitement. The door at Leo's was always opening and closing as one soldier went back to his unit and another came in. The word had gone out that a good time was being had at the house on the corner and it didn't take much persuading to get an *Ami* to come in for food and drink and to talk to the girls.

They stacked their guns in the corner and emptied their pockets. Out of the windbreakers came chewing gum, cigarettes, chocolate, and ration cans of army food, big cans full of meat. Their generosity was amazing, whatever they had they shared with us, and Leo's mother and sister were constantly peeling potatoes and cooking them with the chipped beef or corned beef hash the *Amis* brought. We hadn't seen meat like that in so long—meat and the beautiful brown fat in the pan poured over the potatoes—it was a meal for royalty. The only thing the *Amis* didn't have was liquor, but there was no shortage of brandy or Bols gin. Everyone was breaking out bottles they'd saved for years, and nothing made them happier than to pour out a drink for an *Ami*. We were all smoking American cigarettes and the air was thick with smoke. There seemed to be an endless supply of everything and these big soldiers were so unlike anyone we'd ever seen, just dishing out whatever they had, that we couldn't do enough for them. The women would have washed their clothes on the spot, if they'd asked. They were obviously enjoying themselves too. But several times, in the midst of the laughing, talking, eating and drinking, I would notice a soldier's eyes, as if he was thinking. For the moment he was around civilized people, in a working man's home with women and young girls and children, being treated like a hero, but tomorrow he might have to fight. When I saw the soldiers' guns all propped in the corner of the kitchen I wanted to grab one and run out and fight the Germans myself.

"Josko! You're alive!"

"Vwadzu! You are too!"

It was the first thing everyone said when they saw you, 'You're alive!' Vwadzu had been to trade school and was working, he made it all through the war and so did his brothers, Taduscz and Januscz never got picked up either. When Vwadzu's parents saw me they started crying and gave me a hug.

"You're alive, you're alive!" Mr. Kamic kept saying in Polish. "How was it, Josko?"

"Bad," I said. "It was bad."

Mr. Kamic cursed the Huns, and then another Polish coal miner who lived in Frankstraat came over with an *Ami* he'd found who was also Polish. The soldier was a big blond guy with a Lugar in a holster strapped to his shoulder and Mr. Kamic couldn't believe it when the *Ami* greeted him in his own language.

"He speaks Polish! The *Ami* speaks Polish!"

"I told you! I told you!" the other coal miner shouted.

"Where were you born?" Mr. Kamic asked him.

"America," the *Ami* said. "But my parents came from Poland, both of them. Czestochowa." He got out his wallet and showed pictures of his family.

"Where was this taken?" Mr. Kamic said. "In America?" He still didn't believe it.

"That's my house. That's where we live," the *Ami* said.

"You have a house? Your family's rich?"

The *Ami* laughed and said his father worked in a factory. It was incredible to Mr. Kamic, to all of us. He asked the *Ami* if his father was big, like him.

The soldier pointed at a guy in the photo, a short, older man Mr. Kamic's age, and said, "No, he's like you!" and everyone laughed.

Other *Amis* had pictures of their families too, with houses and refrigerators, even cars, and seeing the way they lived was just like everything else about them that first day. It was all something we had never seen before, something we had never imagined. People were packed in the house till past midnight, eating and drinking what was left of the Bols. We couldn't get enough of it all. The *Ami* who spoke Polish could communicate, but someone was always saying, "Where's Josko? Get Josko!" because I could speak English enough to make myself understood and translate for the *Amis*. We ate, we drank, we sang and laughed, we smoked cigarettes, we cursed the Germans and said whatever we wanted. We were free. The most important thing was seeing the people you thought you would never see again, and for me there were many but especially Vwadzu and his family. Mrs. Kamic had always fed me and the way Mr. Kamic talked and laughed with his sons, and the way he treated me too, that was what a father should be.

"Are you going to go home, Josko?" Mrs. Kamic asked. "I saw Dragitsa. She said someone told her they'd seen you."

I shook my head. I wasn't going to #22. They already knew I was around. When Leo and I first came back to Frankstraat after we escaped from the camp, I learned from his mother that Gizela was alive. She had survived all the time I was gone, but she hadn't recovered. My aunt had told Mrs. Timmerman that she needed lard to feed Gizela, that she wouldn't get well without it.

It was on my mind while I was hiding out in the water tower. Several times I climbed down early in the morning before it got light and went through the orchard and across the field to Frankstraat. I knew there would be dogs at the dump near the cesspool at the bottom of the street. I'd go out when it was still dark and catch one, strangle it, and leave the carcass by the house. My aunt would know what it was for. She would know I had left it for Gizela. She'd boil the carcass and render the fat, and give it to Gizela to help her get well. So they knew I was back. But there was no way I was going to go home.

The next morning after the liberation and the party at Timmerman's, I went out to see what was going on. It was a beautiful, warm day, and I wanted to meet an *Ami* and see if I could get something to eat. They were right there in the field across the road from Leo's, and I could hear shooting from the direction of Heerlerheide. The town apparently hadn't been liberated yet, so I headed down toward the Dutch Alp where the road to Heerlerheide turned off the main one. A half-track was parked by the slag pile, the "mountain", and I walked over to it, stepped up on the tire, and pulled myself up to look inside. Two soldiers were sitting in the cab. I said, "Hello," in English.

"Hi, kid," the driver said. "You got a sister?"

"Yes," I said.

The *Ami* laughed and started the engine. I jumped off the tire and they pulled out. All the vehicles lined up along the road were starting their engines and moving out. I ran along behind as the trucks and half-tracks drove a short ways, then they turned off the road and drove into the wheat field. Beyond was the orchard and water tower. They stopped when they got to the orchard and a bunch of soldiers jumped out and started digging foxholes. They had hardly moved at all and now they were digging in again. I strolled around watching them for a few minutes and stopped near a guy spading up the earth with a small shovel. He had his back to me and all I could see was how big he was, just huge, even bigger than most of the others. He was digging without much enthusiasm, so after watching him for a couple minutes, I said, "Hi," to get his attention.

The *Ami* turned around and I almost took off. Not only was he so big, he had the ruggedest, ugliest face I'd ever seen. It was one big wrinkle, a frown that ran from his chin all the way up under his helmet. He said, "Hi," back to me, although it was more of a grunt, then went on digging.

He hadn't told me to leave, so I stood there, watching him. I could see he wasn't too eager to dig a foxhole, so after he turned up a few more clods, I said, "Can I dig for you?"

He stopped again and turned around. The ugly, wrinkled face looked at me like I was crazy.

"You want to dig?"

"Yes," I said.

He stood there staring at me. Then he gave me the shovel.

"Okay."

I started throwing up dirt, digging like I really meant it. I heaved the soil on one side of the hole, all in the direction of what I thought was the "front" and packed it down with the back of the shovel. This was my interpretation of how to build a defense perimeter. As I dug I kept glancing at the *Ami* to see if he approved of what I was doing, and it must have amused him because every time I saw his grubby face it had a lopsided grin on it. He let me dig a while before he said, "Do you have a sister?"

"Yes," I said.

"What's her name?"

"Dragitsa."

"You have any schnapps?"

"No." The night before I'd learned that 'schnapps' was what the *Amis* called liquor of any kind, whatever it was.

He watched me dig a while longer and when I'd gotten as far down as the length of the shovel, he said, "Are you hungry?"

"Yes," I said.

He picked up a small canvas bag, reached in and produced a loaf of bread. He tore off a piece and handed it to me. I took the bread and stared at it. It was the whitest thing I'd ever seen in my life, so white I couldn't believe it was bread. I took a bite and had another surprise. It was so sweet it didn't taste like bread at all. It tasted like cake.

The *Ami* reached in the canvas bag again and took out some butter. He spread some on his bread, and said, "You want butter?"

I shook my head no. I had never refused butter in my life, but you don't put butter on cake!

Now from the canvas bag he produced a can and opened it.

"Ham and eggs?" he said.

I said yes and he handed the can to me. He put the hunk of butter back in his bag and watched me eat, all the time with that rough, crooked grin.

"You dig pretty good," he said. "What kind of work do you do?"

"I don't have a job," I said. "I escaped from a camp in Germany. I worked in a coal mine."

"What's your name?"

"Josko."

"Is that a nickname?"

I didn't understand the word in English so I just shook my head.

"Is that the name you were given at birth?" he asked.

"No. My name at birth was Josef," I said.

"'Yosef?" he said. "We call that 'Joe.'"

"Like Joe Louis?" I said.

The *Ami* nodded.

I had the same name as Joe Louis! He was the great American hero who knocked out Max Schmeling before the war. The day after the fight, when I walked to school all the German kids had their heads hanging! Joe Louis was America itself!

"My name is John," he said with the big crooked grin. He took out a pack of cigarettes, Philip Morris, in a brown wrapper. "You smoke?"

"Yes."

He gave me a cigarette and lit it.

"Here." He handed me the rest of the pack and stepped down in the hole I'd dug for him. I put the cigarettes in my pocket, thinking that was my pay for helping him, the cigarettes and the food. While I smoked he started digging again.

"You speak pretty good English," he said.

"I learned it in school."

"Are you Dutch?"

"No, Yugoslav."

"You speak Yugoslav?"

"Yes. Also Dutch and German."

"German, too, huh? You live around here?"

"No. I don't have a place to live. The house was bombed and the family was killed."

John had stopped digging again, taking a break.

"Where are you from?" I asked him.

"Oklahoma," he said.

The name meant nothing to me. I'd never heard it in school or the movies.

He dug a while longer and had almost finished the foxhole when a jeep came bouncing across the field toward the orchard. As it went by us, the *Ami* riding beside the driver shouted, "Get a move on, soldier! We're pulling out!"

The jeep kept going and John said something back to the guy, most likely a curse, after the effort of digging the hole. He folded up his shovel and picked up his pack and canvas bag. The activity was starting up all around us in the orchard as they got ready to move out again, but just like yesterday when I first saw the *Amis* in the square, I noticed the way John and the other soldiers behaved because it was so different from the Germans. When the officer gave him an order, John was cool, he didn't jump, and when he did move he acted like the order was an imposition on him. When one of the trucks drove near us, John jogged over and threw his pack in the back. Behind it was a half-track moving slowly. As it drove by, John handed his rifle to one of the soldiers, jumped up, and climbed in back with the others. All around as the half-track crawled slowly away were the sounds of people shouting orders, engines revving, tanks rumbling along, and in the distance artillery was blasting. This was it, the front line, there was no one between them and the Germans. They were advancing to fight.

I started walking along behind the half-track to follow the action because everyone was moving out, clearing the orchard. John was sitting at the end of the half-track and when he glanced back and saw me he waved, just a quick flip of his hand as if to acknowledge me, to say goodbye.

All the vehicles were slowly forming into a column, churning up dust as they advanced back across the field to the road. Over the roar of the engines I heard someone shouting in the back of the half-track. “Hold it! Hey, hold it!”

The half-track stopped short and John shouted:

“Hey—you want to come with us?”

This took me by surprise and I just stood there, not sure what he had said.

John waved his hand and shouted, “Come on! You want to come?”

As soon as I realized what he was saying, there was no question. It was automatic.

I ran to the half-track, John reached down, grabbed my hand and helped pull me up.

“Okay!” someone shouted. “Let’s go!”

We started moving.

“You want to come with us?” John asked me again.

I nodded. The noise of all the mobilizing soldiers and vehicles was roaring all around.

“Is he coming?” someone shouted.

John looked at me standing in the back of the half-track as it bounced along. “Yeah, he says so.”

“Well, get him some clothes,” the guy said.

I was still wearing the sport coat and pants and shoes I’d stolen from the house in Germany after we got out of the camp.

“Hey, Jimmy,” John shouted up toward the front. “Give this kid something to wear.”

A guy near the front grabbed one of the duffel bags on the floor, opened it and started pulling out clothes. On the benches in the half-track eighteen or twenty soldiers in helmets sat chewing gum and smoking, just riding along, holding their rifles. A couple of them nodded at me.

I looked out the back at the orchard and water tower slowly moving away and it hit me—the roar, the trucks and tanks, the artillery firing, jeeps rushing by with commanders yelling—I was riding along in a column of American soldiers.

"Here." John handed me the clothes the guy called Jimmy had tossed down. "Put 'em on."

I took off my jacket, shirt and pants, and standing there in the middle of the half-track with all the *Amis* watching me, I pulled on the combat fatigues. The guy who'd given them to me was the only one anywhere near my size and even his were too big. The guy also had an extra pair of boots. They gave me a duffel bag to stuff my own clothes in and John handed me a helmet. Then he reached under the bench and took out a rifle, much smaller than his. "This is a carbine," he said. "Ever fire one of these?"

"No."

He handed it to me. "The one thing you've got to know is the safety, this thing right here. When you push it out to your left it's on safety, so it won't shoot. Push it to your right and you're ready to fire."

I nodded.

"Here's the clip with the ammo. You push it in with your thumb, like this. Got it?"

I nodded again.

"Sit down," John said.

That was my basic training.

I sat down on the bench in my new clothes and helmet with the rifle between my knees. The air was full of dust and exhaust from all the vehicles in the column, and over the roar of engines I could hear the artillery shells firing up ahead, a little louder. Slouched in the half-track, riding along, the men were quiet, subdued, their faces were stone, perhaps because of what they had been through. But looking at them I saw nothing of fear, no defeat in their eyes. It was so different from what I'd seen in the camp. For two years I'd lived with beaten—just brutally beaten people—with fear in their eyes, in their faces, in the way they walked. These men were different, and riding along with them—although this was the front line, we were it—I felt things were going to be all right.

The half-track suddenly lurched to the right and stopped.

"Everybody out!" somebody shouted.

I jumped out with the rest. The tanks had pulled off to the other side of the road and I could hear machine gun and rifle fire. Another huge soldier was shouting orders to everyone, telling them what positions to take. He ignored me, so I followed John as he ran toward a garbage dump. Bullets were flying around us but there was no panic, John just jogged in a natural rhythm, hunched over and keeping his head down as he looked around. The dump was a concrete basin six or eight meters long and a meter deep, like the one at the bottom of Frankstraat. We crouched behind it and John started shooting at the Germans trying to advance across a field. When I saw him shooting, I set the carbine on the lip of the basin, aimed at a *Mof* and pulled the trigger. I heard a CRACK! and felt it recoil against my shoulder. I had a gun, I was shooting at them! I emptied the clip. The Germans were already turning back and as they retreated the tanks rolled across the field after them. It had been a halfhearted attack and no one seemed very excited or even relieved when the order was passed along: "Get back in the half-track." As I followed John back, I saw a couple *Amis* pushing two captured Germans along with their hands on their heads.

The squad gathered around the half-track and lit cigarettes, waiting to see whether they were going to move out right away or not. I kept my eye on the Germans. They were leaning against a truck with their hands still up on their heads. They were older soldiers who had obviously seen a lot, like the guys I was with, but nevertheless they looked scared. Later, after I'd seen more myself, I understood why. When a lot of men gave up—ten, twenty, thirty—it was different, something had to be done with them. But just two guys...what happened if the enemy started pushing back, what would be done with them? They would become a burden. The *Moffen* were thinking that might happen to them, they might just disappear instead of being sent back as prisoners. But while I was watching, one of the *Amis* who had captured them walked up, opened a K-ration container and pulled off the paper. He took out a small pack of cigarettes and offered them one. The other *Amis* were watching too. The *Moffen* took the cigarettes and the *Ami* lit them. Another soldier came over and said, "Are you hungry?" The *Mof* shook his head, he didn't understand, and the *Ami* made a gesture of eating and the *Mof* said, *"Ja, ja."* The *Ami* opened a larger ration can and handed it to them and also gave them his canteen. They took the food and water and didn't look as scared. They weren't going to be fed and then shot.

A few minutes before when I was crouched behind the garbage dump with John, all I could think was, kill a German, kill a German, kill a fucking Hun. I wanted revenge. Now I felt something else. The *Amis* were rough—they shouted at each other and cursed in strange words I'd never heard before—but the sound wasn't harsh. They weren't hysterical like the *Sondermeldungen* on the radio, or the guards in the camp. They didn't have the insane, violent shriek of all the Huns who'd had me under their thumb. There was something about the *Amis*, almost a gentleness, and seeing how they treated the *Moffen* took some of the anger out of me.

When we climbed back on the half-track we rode to the edge of a town a few kilometers from Heerlerheide, which by now had been liberated. We didn't see any more action though I could hear guns firing and shells blasting all afternoon. We spent a lot of time sitting in the half-track waiting for the column to move and by late afternoon the tanks had taken up positions and the word came down that we weren't going any further. The half-track drove into the yard of an elementary school, a school I recognized because *Mijnheer* Van Dijk's father was the headmaster. It was vacant now and we were going to sleep in the schoolhouse. We piled out and the first sergeant came around checking to see who'd been hit or wounded during the fighting. I didn't know rank yet but I was starting to notice how many stripes a guy had on his sleeve, and he talked to one guy after another. I was standing with some of the soldiers when all of a sudden he saw me and stopped short. He stood there staring at this anemic-looking soldier with a carbine who'd just appeared in his unit and said, "Who the hell're *you?"*

I pointed at John who was standing a short ways away talking to another soldier. "John picked me up," I said. "Another man gave me clothes."

All day no one had challenged my being with the soldiers, they just seemed to accept it, but the sergeant was obviously someone in authority who could kick me out. The other thing that scared me about him was his size. John was big but he was lean, built like a cowboy. This brute not only had a mean face, but a big gut hanging over his belt—a good 250 pounds of indignation. He was the biggest man I'd ever seen.

"So what the hell're you doing with that *rifle?"* he said and grabbed it from me. He thought I might be some crazy Hitler Youth kid who walked into camp to shoot it out with the *Amis*.

"John gave me the gun," I said. "He asked me to come with him."

"Roper!" the sergeant shouted.

John turned around and ambled over, taking his time.

"Who the fuck is this? You start recruiting for Uncle Sam?"

John gave the sergeant his crooked grin.

"He's okay, Sarge. He helped me dig a foxhole this morning. I picked him up because he speaks English—he said he speaks German too. I figured we could use him when we get into Germany one of these days."

"He speaks Kraut?"

"That's what he told me," John said.

The sergeant looked at me with his frown. "You speak German?"

"Yes," I said.

"Hey, Gingold!" the sergeant shouted over his shoulder.

Another soldier, an older guy, came trotting over and the sergeant said, "This kid says he speaks German. Check him out."

The *Ami* named Gingold started speaking to me in the weirdest German I'd ever heard. It turned out he was speaking Yiddish and although I'd never heard anything like it I got the idea he was asking me where I lived because it sounded like *"Wo wohnst du?"*

I said, in German, "I came from Holland."

"Why do you want to come with us?" he said. Once again the language sounded all screwed up, but I got the meaning.

"I have no place to live," I said. "I'm alone. My family is dead. They were killed in the war."

"He says he's all alone, his family got killed," Gingold said to the sergeant. "The fucking kid wants to come with us."

I understood what Gingold said to the sergeant in English better than his crazy German, and could have told them the same thing in English myself.

"He's crazy," the sergeant said, frowning at me.

"Du bist verrückt!" Gingold said to me. "You're crazy! Go home!"

"No, I'm not crazy," I said in German. "I want to come with you."

Just for the food alone I wanted to stay with them—bread like cake and cans of meat—if they only fed me I was being overpaid, besides the cigarettes, clothes, boots, a helmet, a rifle. If I stayed with them I could eat and fight the Huns too. What else did I want?

"He says he's not crazy," Gingold told the sergeant.

"Can he really speak German?" the sergeant asked Gingold.

"Ich spreche sehr gut Deutsch," I said to Gingold before he could answer the sergeant. I wanted Gingold to know I spoke German very well because whatever he was speaking sounded so strange I didn't know if he knew what good German was.

Gingold looked at me. "Yeah, he speaks it pretty good," he told the sergeant.

The sergeant stood there thinking.

"What the hell. We'll find something for him to do."

He frowned at me.

"What's your name, kid?"

"His name's Joe," John told him. "Little Joe."

"Okay, Little Joe. Have you got a mess kit?"

"No." I didn't know what that was.

He shouted at someone else who turned out to be the unit's supply sergeant. Another big guy came over.

"Get Little Joe a mess kit and some underwear and whatever else he needs."

"Sure. Who the hell is he?"

"Some kid Roper picked up this morning. He speaks German."

"Oh, yeah?" the supply sergeant said, like it was a joke. *"Sprechen Sie Deutsch?"*

"Ja," I said. *"Ich spreche Deutsche."*

"Hey, he's a Kraut-talker," the guy laughed. "Come on, we'll get you a mess kit and fart sack."

I was hearing a lot of words they never taught us in school.

As I followed the supply sergeant, I heard the first sergeant say to John, "Watch his ass, Roper. Make sure he doesn't shoot one of us."

After I got a mess kit and fart sack, a sleeping bag, John said, "Come on, Little Joe. We're going to eat."

I followed him over to where they had set up the field kitchen and we got in line. The guys around us talked and rattled their mess kits, the atmosphere was loose and relaxed, but it was all strange to me. As the line moved up, the first thing I came to was what looked like a big trash can filled with steaming hot water with gas burners under them. I watched John and when he dipped his mess kit in the can and shook off the water, I did the same. Next we came to a table where the food was laid out, big pots of beans and trays a good meter long and almost half a meter deep filled with elbow macaroni with some kind of yellow stuff on top. Then piles of bread, the same incredible white bread John had given me in the morning. When my turn came, I stuck out my kit and the cook who was serving—he was another giant, they were all huge—dipped the spoon into the macaroni and dropped a mound in my kit. I started to move down the table when he scooped out another spoonful and said, "You want another?"

"Yes," I said, and nodded at the cook. He dropped it in my kit and I heard a couple of the guys chuckling behind me in line. The next guy spooned a peach dessert in the other side of the mess kit and I came to the bread. I still couldn't believe how white it was and now I had another surprise. It was sliced into neat ovals, all the same thickness. I'd never seen bread in slices and there was so much of it I grabbed four slices. No one said a thing. Beside the bread was butter in cubes and I took a fistful, then I got coffee. I had never seen so much food and I sat down beside John on one of the long benches they had set up and started eating. I didn't taste a thing, I just shoveled it in and stuffed my gut as fast as I could. When I finished John said, "Little Joe, you want seconds?"

I didn't know what that was.

"You want more?" he said. The other guys were watching and smiling.

"Can I have more?" They gave me so much the first time I couldn't believe they would let me have more.

"Sure, go on back," John said. "We got plenty of that shit."

I went back to the food table and guys were laughing at the cook, making jokes because someone actually wanted his food.

When I came back with two more spoonfuls of macaroni and sat down again, I realized I couldn't eat as fast. I'd already eaten way above my stomach's capacity but I was afraid if I didn't eat it all I'd never get any more. If I wasted their food they might throw me out. Eating more slowly I started tasting the food and realized the yellow stuff on top was cheese—something I was allergic to. I got it all down even though I knew I was going to be sick. So I made two butter sandwiches with the bread and put them in my pocket. Sure enough, it wasn't long before I got a fierce headache and all the macaroni came up. Then I was fine, just hungry again, so I ate my butter sandwiches.

After supper the guys sat around talking and smoking and some of the others in the unit came over and asked John—"Big John" they called him—who I was. "This is Little Joe," he told them. "He speaks Kraut."

"A Kraut-talker, huh? Who fixed him up with the clothes?"

"Jimmy did." Jimmy was the squirt, the only little guy in the outfit. Everyone else in the squad was big, big and friendly. No one seemed surprised John had picked up some kid—except me—and it was like the night before at the Timmerman's, I was discovering the generosity of the Americans. For them it was nothing to adopt someone into their outfit, even the sergeant in charge didn't mind, and the other guys just nodded and said hello. If Big John said I was okay, that was enough.

After I got sick and ate my sandwiches, John and I sat together smoking cigarettes and he asked me a few questions about where I'd come from and what I'd done during the war. I told him I'd been in a labor camp in Germany for two years, working in a coal mine.

"Was it hard work?" John wanted to know.

"Very hard work," I said. "Many died."

"How'd you get out?"

"I climbed over a fence while the English were bombing. I came back to Holland to wait for the Americans. I was hiding."

"How long have you been hiding?"

"Three weeks," I said.

"Have you got any relations?"

I didn't know what he meant.

"You got any family?" John said.

"In Yugoslavia, if they're alive."

"So you're all alone."

"Yes."

Big John put his hand on my neck and gave me a squeeze.

"Okay, Little Joe. You stick with me, I'll take care of you."

He gave me that ugly, lopsided grin and stood up.

"Come on. Let's go find your fart sack and grab a place to sleep."

Chapter 7
Border Between Holland and Germany; Fall 1944

The Americans who picked me up were in the 29th Division, Ninth Army, and had been fighting their way across France, through Belgium and Holland. Now they were right at the border of Germany. They were seasoned soldiers, not new recruits. Some of them, like Roper, had been in North Africa, then sent to Europe where he'd been wounded, recovered, and sent back to the front. The D-Day invasion at Normandy had been the previous June and the *Amis* needed all the men they could get—not that they were looking to pick up kids who weren't seventeen yet—but a lot of guys who got wounded, if it wasn't too bad, they were treated and then sent back to whatever units needed reinforcements.

My first impression of John B. Roper was right. If I'd seen his face first, before I'd spoken to him in the field, I'd have been too scared to say anything because he was so ugly he was intimidating. But when he said he'd take care of me, I believed it, and that's the kind of man he was. If you wanted a friend, you wanted Roper; if you wanted an enemy, you did not want him to be Roper. He was no good, a professional private. Whenever he got promoted he fucked up and got busted again. The first sergeant, Hannah, was different. He looked like a big, fat cop, a brute with a gut, and he had a mean face, always with a frown, but he was good-natured. The company commander was Captain Hardesty, the heavyweight boxing champion of the Hawaiian Territory before the war. The first time I saw him was the next morning when we were bivouacked at the schoolhouse and he came out to stand reveille. Three kilometers away I could hear shooting at the front and the captain is standing there wearing a long velvet robe and slippers, his elbows resting on his gut, smoking a cigarette. A real hotshot. The unit also had a Polish guy, a German guy whose parents had been born in Germany, and an Indian called "Chief." Whoever said don't give an Indian a drink must have been thinking about Chief because when he was drunk he was crazy.

My first morning with the American army was a repeat of the night before. They served powdered eggs, powdered milk, pancakes, coffee—plenty of real coffee—and to the *Amis*, this stuff was terrible, but to me it was the first time in my life I was able to eat as much as I wanted. Just for the food I felt like I was being overpaid. But Sergeant Hannah said, "We can get Little Joe his cigarette ration, but how the hell are we going to pay him?" You got seven packs a week free—cigarettes, chewing gum, toothpaste, soap, it was all free from the quartermaster—and now they were worried about paying me! I'd never had money in my life. What was I going to do with it, anyway? Just feed me!

I was also picking up some of the GI lingo, nothing like the proper English we heard in school. When the guys referred to an officer, it was "the fucking captain," "the fucking sergeant," although I quickly learned when you talked to an officer you left out the other word. But everything about the Americans was just like the first time I saw them in the square when they liberated Heerlen. They were a kind of person I had never seen before.

The unit's base of operation was the schoolhouse in Heerlen. We made forays into Germany and then fell back to the schoolhouse, just as other units might go forward and then fall back to Maastricht or Venlo. After the first three nights there we got orders to go east, toward Germany. The column moved out and on the way, while we were still in Holland, someone started shooting at us from a building. He was in the cellar, shooting at ground level, and it wasn't just potshots, he really meant to hit someone. I went with Roper and Hannah and a couple others when they surrounded the building. Two guys went in from the rear and they brought him out alive. Instead of a soldier, it was a guy wearing a blue suit with a white, sleeveless undershirt. He was dressed as a civilian but he was *SS*, with an "E" tattooed under his upper arm. All the *SS* had this tattoo. Under the Geneva Convention you can wear civilian clothes as long as you have a uniform on underneath or wear some military insignia. Otherwise, you can be killed by firing squad because you're considered a spy, so this guy got a very rough treatment. Roper was pissed off and instead of shooting him, he put the guy on the hood of a jeep, mounted him behind the wire cutter, and we took him for a joyride. The Americans drove their jeeps with the windshield folded down, so the Germans would string a single strand of wire across the road so when you hit it, the wire cut off your head. You'd lose the driver and maybe the lieutenant or captain riding in front beside him. To prevent this the Americans welded a piece of iron on the bumper of the jeep, standing up about a meter, three feet high, with the top angled forward so when the wire hit it, it would snap off. John B. made the *SS* guy sit on the hood and straddle the wire cutter with his hands behind his head. He wasn't allowed to hold on. Every time John B. made a fast turn, the guy would reach for the wirecutter and John B. would rap him with his rifle and say, "Don't do that! Little Joe, tell the fucking Kraut he's not supposed to do that!" I yelled at the guy, cursed him. "*Dreckiger Deutscher!*" I was in my glory because this guy was *SS*, a German everyone hated, a real arrogant sonofabitch. When the jeep turned and he didn't hold on, he fell off. John B. stopped the jeep, and told me to tell him to get the fuck back on. I was giving orders to the *SS*. Finally when he fell off, John B. just drove over him.

American troops had already crossed the border before we did, so we weren't the first—and no one in the unit really said anything about it—but after all the fighting they'd been through

I'm sure it meant something to the guys I was with to enter Germany.

The scene we came on looked like a real mess, a battlefield. You could compare it to an untidy room. The Germans had put on a very strong resistance along the border and the roads and fields were in total chaos, with junk and material all over the place. Bodies were lying everywhere, dead German soldiers, busted tanks and vehicles. The Germans still used horses to haul supplies and quite a few had been killed, the big carcasses torn open by shells. Things looked dismal even in the villages with buildings shattered and blown up because the American artillery would blast a village before the tanks shot their way in. The guys in the unit talked about how it was different from when they'd been in Holland or Belgium. There they had been very reluctant to shoot up the towns because they didn't want to hurt the people. It was like the reception they got in Heerlen, whenever they came into a new town everyone would turn out and go crazy. So they didn't blast their way in. If a Kraut was shooting at them from a building, they'd send in the infantry to flush him out, like we'd done with the *SS* guy just before we crossed the border. But in Germany, if someone shot from a house and there was a tank close by, they would just call up on the walkie-talkie and say, "Fire from your position, one o'clock, two o'clock," whatever it was, describe the building, the tank would fire the howitzer, POFFF!, then run right into the house, and back out. The guys felt this was the way it should be, fuck the civilians and houses, just blow 'em up. I did too. I felt that now the Germans were getting a taste, they're finding out what it's like.

When we came back through the same area a few days later, the Corps of Engineers had cleared the roads and checked out minefields, and the dead bodies had been cleared out. The animals had been buried by German farmers because everyone was worried about an epidemic. We made several forays from the schoolhouse into Germany before advancing for the battle of Aachen. It was near the end of October when we crossed the border and continued to a town called Alsdorf. It had already been taken and the German civilians were walking around very cautiously. Aachen, our objective, was ten or fifteen kilometers south and Sergeant Hannah told us it was going to be a tough place to take. It was important symbolically, an old German city with the oldest cathedral, the place where Charlemagne had been crowned emperor, and the Germans had pulled out the regular army, the *Wehrmacht*, and had brought in *Waffen SS* to defend the city. They were not going to just give it up. As we turned south, bypassing Herzogenrath and Kohlscheid to the east, we came to where the trolley line ran almost directly alongside the Dutch border, but unlike when Leo and I escaped from the camp, now there was no fence. The Americans had rolled their tanks over it. A lot of units were moving into position to attack Aachen and all the roads we passed were jammed with equipment. Supplies and ammunition were stacked up everywhere, on main roads, dirt roads, in the fields. The effort by the American army was incredible. All the shells, the trucks, the food were American, and after the war when I talked to Germans they couldn't believe it either. "Only the Americans brought a bridge with them!" they said. Another surprise was that all the truck driving, loading and unloading of supplies was done by black units. I had never seen a black person so to me they all looked the same. I couldn't figure out how they could tell one man from another.

It was a Wednesday when we got close to Aachen. We stopped on a hill and were told to dig in and establish our position until word came to move out and take the city. We heard that the German *SS* were running convoys from the east at great effort through the American lines in order to supply the encircled units in Aachen. Some succeeded, some didn't, they were battling it out. That was on the opposite side from where we were but the firing was heavy. In back of us artillery units were setting up long-barreled guns, fifteen feet long, two feet in diameter. The shell stood four feet high and it was a bomb—the powder was in bags, they weren't encased—and the guns were pulled by tractors so big they would dwarf a farm tractor. The guns were such monsters that when they went off all hell broke loose, the ground shook, roofs rattled. By Thursday afternoon we were all dug in and if they tried to break out we were ready for them. Contact was made with the defenders of Aachen and a bulletin was broadcast over the radio to the Germans—if the city doesn't surrender by three o'clock Saturday, we're coming in. It wasn't a secret, the American command was giving them a chance to surrender. Of course these fuckers aren't going to surrender. No way. In the meantime, sort of polite shelling went on, guys just checking to see if their guns worked.

Saturday was a cold, raw, nasty, overcast day. The end of October. Everyone was bored and edgy waiting for the deadline. In the morning one of the guys in the unit found a bunch of trolley cars near the top of the hill where we were dug in. "What if we run one down into Aachen and see how far it would go?" This idea created some interest. The trolley tracks were less than a hundred yards away and all around us were piles of ammunition the Germans had left behind, and someone else said, "Let's load it up and send it down to the Krauts." We filled it up with all kinds of ammunition and shells, including a lot of what the Germans called the *Panzerfaust*. It was the equivalent of an American bazooka, a pipe with a big knob on it. *Panzer* is tank, *faust* is fist, it was the fist to knock out the tank. While we were loading it up Captain Hardesty came around. "What the hell are you guys doing?"

"We thought we'd run this down into Aachen."

"Who's going to drive?"

"Nobody. We'll let it coast down the hill."

"You ought to wait till three o'clock," he said. "Just in case."

"Shit, by that time all the Krauts will be tucked in."

None of us thought they'd surrender.

"We can always load up another one before three."

"Okay," the captain said. "Go ahead."

Roper backed a jeep up the hill alongside the tracks, nudged the bumper against the coupling on the back of the trolley and pushed it up over the top of the hill. It started down and we all stood there watching it pick up momentum. It kept rolling and rolling until it went out of sight and we still didn't hear a thing. No crash, nothing. We figured it must be still on the tracks, just rolling along. About half a minute later there was a god-awful explosion in downtown Aachen and a cloud of smoke rose over the city. So much for honoring the deadline, which made no difference anyway, because the Germans had no intention of surrendering. At 2:58 the planes started coming in and at three the air was filled with bombers and the artillery let loose, blasting and blasting. Aachen had not been bombed too severely before but now they were getting it. We just held our position and watched while everything shook. The shelling and bombing went on through the night until Sunday. About ten o'clock in the morning we were told to go in.

We followed the tanks. It was fierce. It hadn't been long since the Allies entered German soil and the Germans still felt very strongly—especially the *SS*—that they could defeat the invaders. They were fighting for the high Fatherland and were ready to slug it out toe to toe, and that's what it was, street fighting, house to house. I had become a soldier in very short order. I could understand orders, the field and platoon sergeant and company commander were very efficient, and I was treated as part of the unit. "Little Joe, get over behind that rock and cover the three windows on the right." Or, "Cover the second story." The honeymoon was over. Little Joe had to earn his three square meals and seven packs of cigarettes a week. I still carried the carbine because I could handle it, but when we went into Aachen we all had carbines since it was designed for street fighting. It was much easier to load because you just dropped the clip and put another one in, and the clip held thirty-five bullets and the M-1 only had eight.

As we followed the tanks, three or four guys walked on one side of the street and another five guys on the other. We moved forward, running from doorway to doorway until we drew fire, then we dispersed, got under cover, and the sergeant pinpointed where the fire came from. The four guys on the other side of the street would pin down wherever the fire was coming from while one of us moved up and threw a concussion grenade at the house. After it exploded, we went in. We didn't knock on the door and say, "Come out!", we blasted the place and went in to kill. We searched the house top to bottom, down to the cellar. That's where the civilians were always hiding if there were any. They were petrified but we left them alone and told them to stay there while we looked for anyone who might be an army or an *SS* man.

The others in the squad stayed outside to cover us until we came out. Sometimes it took a few minutes, sometimes it took an hour. If it got too tough to handle, the platoon sergeant called for a tank and if they could spare one it would barrel up and blast the building with five or six shells.

The battle for Aachen took over a week and every day was much the same, going house to house, building to building. None of the guys in my unit got killed but several of them, including the supply sergeant, got pretty badly shot up and were sent back to field hospitals. The prisoners we took were all *SS*, and they were young, eighteen, twenty years old, very fresh but very well trained, perhaps because their officers were older. When the fighting was over we moved into a row of middle-class apartment houses. The civilians had been told by German propaganda that once the enemy got into Germany we were going to kill all the women and children so they were terrified but all we did was throw them out. We kept some of the women and told them to make our beds, but we didn't touch them, there was no raping, we just made ourselves comfortable.

I was treated like the rest of the guys. I had to do a job and was always under the close scrutiny of the squad leader just like everyone else. Some of the assurance of the others rubbed off on me but during the fighting I had no fear. I walked right out and shot at the Germans. What I felt was—I had always known I would escape from my father, I would get away from him, and when I did nothing could touch me. It was like in the labor camp; I knew I would survive, and fighting with the American soldiers was the same thing. I had no fear. Because the worst thing for me was not labor camp, or war, it was my father, and as bad as he had treated me physically, what I hated him for even more was what he said, the words he used to curse me, and my mother, the contempt he reminded me of every day.

But what I felt for myself because of my father or the Germans was totally different from what I felt about what had been done to my sister. Even the incident of shooting every third man in the labor camp was nothing compared to what I felt for Gizela. Because of her, I felt such a brutality that I had absolutely no sorrow about killing the Germans. For the Americans I was with, although they had been through a lot, it was different, and they taught me something. They taught me something about compassion. When we were in Aachen mopping up, after the city was taken, I walked into a house. There was a young German woman inside and I said to her, "I want a glass of water." She got me the water and when she gave me the glass she said, "Is that all you want?" The way she said it was—she was insinuating something, as if she was saying I was that kind of person and she expected me to do something to her—and I knew damn well what I wanted to do to the Germans, I wanted to kill them and fuck them—so when she asked me the question it made me so angry I totally lost my head. All the anger and bitterness I felt about Gizela came out, because this was a young woman my sister's age, my sister who was still lying in bed dying. I beat the young German woman. I beat her to a pulp. I was going to kill her. Sergeant Hannah heard the screaming and came in and grabbed me by the back of the neck. He dragged me out and threw me in the half-track. After that they watched me. They watched me a lot.

After the battle of Aachen, we were relieved and fell back to our base at the schoolhouse in Heerlen. We went to visit some of the guys who'd been wounded during the fighting, like the supply sergeant, and he rejoined the unit after three or four weeks. It was a cold, hard winter and we kept going to the front for a few days or a week of action, then returned to the schoolhouse. The objective now was to cross the Rhine River, but from December into January they were fighting the Battle of the Bulge to the south of us. The Germans had broken through the Allied front and were making a stand, so that delayed the advance toward the east.

One day when we had moved up to the west side of the Rhine I was "volunteered" to go on one of John B.'s extracurricular raids on the Germans. There was some fighting in the area, on the west side of the river, but the German soldiers seemed lost, like their communications got screwed up, because the poor bastards didn't want to fight at all. A few got shot and the others were taken prisoners. In the evening after we ate, Roper and Riley, a wiry, bowlegged guy from the mountains of West Virginia, took a bath in the river. It was cold but they went in bareass, soaping up, splashing, daring the Germans on the other side to shoot them. They did it just for the hell of it, because neither of them was especially hung up on baths. I don't think Roper had had one since Africa. When they came out, Roper said, "Little Joe, we need a Kraut-talker tonight. We're going to see the Krauts in Cologne." I thought he was kidding. We hadn't taken the city yet, the Germans still occupied it, but they were going to float down the river into Cologne!

The river was wide, and it wasn't a stream—that thing flowed—but when Major Nyman showed up I knew they were serious. Nyman was assigned to headquarters but he'd been with Roper in Africa. They both got wounded and sent back to a hospital, and when they recovered they were reassigned to different units. Nyman had a college degree and Roper was a cowboy, but they were buddies, and they were both bad. The two of them would go off together behind the lines and kill Germans, but only officers, and come back with Lugers and P38s to prove it. Tonight they were going down the Rhine to Cologne. We piled in the rubber raft, Hannah and Riley were with us too, and paddled across. They had a map and had figured out how far the current would take us downstream before we landed. I was so naive I thought the idea was to hit the bank, run down a ways with the boat, and show off. Nothing doing. When we landed, they tied up the raft and we went into the city, a good ten blocks. The sky was clear but it was dark, blackout, and they seemed to know what they were doing. Hannah said, "Little Joe, you stay with me." He pointed to a spot a couple blocks up the street and said, "Go up there and wait." Which I did. No one was out but wherever I looked I thought I saw a German soldier. I waited and waited, finally I heard blasting, grenades going off, then German artillery and *Panzers* firing. A few minutes later, Hannah and Roper and Riley and Nyman came running toward me, and now the Germans came out in the street. They didn't even see us. They were expecting an army, not five guys, and we ran right past them, back to the raft. Naturally, when we paddled back across the river we landed in the wrong place and the Americans didn't know who the hell we were. "What's the password?" Nobody knows. They called on the field phone, relayed it to the command post where Jimmy Birdsall answered, the company clerk, and he said, "Who's their commander?" "Nyman." "That's them." When Captain Hardesty found out he was pissed off Nyman didn't take him along. I made three more trips across the Rhine into Cologne and the purpose was always the same, raise hell, which was easy to do because the Germans are great for excitement, screaming and firing their howitzers. They think it's the Ninth Army coming across but it's only four or five guys. Once I saw what they were doing, I filled my pockets with grenades, we all had Tommy guns, and we sprayed them around and lobbed grenades to scare the hell out of the Krauts. They didn't need me, I went along just in case they got in trouble and needed a Kraut-talker. But the only trouble we got in was coming back because after we paddled like hell we invariably wound up in the wrong place. All this to me was an excitement, a great adventure. Not that I would have chosen to do it, but as soon as Riley or Roper said, "Little Joe, we need a Kraut-talker tonight," there was no question of not going. I was earning my food and cigarettes.

Early in March we pulled out and left the schoolhouse in Heerlen for the last time. The Allies had crossed the Rhine a week before and we headed north toward Viersen, a city in the Ruhr industrial area. The fighting wasn't tough although progress was slow because the Germans blew up bridges before they took off. When we reached the outskirts of Viersen we took over a factory where they made gas stoves. It was Sunday, no one was working, and Captain Hardesty made it our command post where we'd stay till we got further orders.

The building was like a warehouse with a concrete floor, not very nice, but at least we were out of the cold and the rain. Across the road was a row of working class homes and someone suggested we throw the Krauts out and let them sleep in the factory but instead we just went into their homes and hauled the mattresses out. They could get them back when we left. While we were unloading the jeeps and trucks in the dark, a German plane flew over strafing. We all put on our helmets and the charge-of-quarters, a guy called Ozzie, jumped on the captain's jeep and started shooting the .30 caliber machine gun at the plane. Tracers were already going up from an anti-aircraft unit but this clown is going to bring the plane down by himself. The captain ran out of the factory yelling, "You stupid sonofabitch! What the hell are you doing? You want that fucking Kraut to lay a bomb on us?"

"He's right there, he's right there!"

"I know he's right there! Let the anti-aircraft nail him. You think you're going to bring him down in the dark with that fucking thing? Knock it off!"

Ozzie was another crazy guy, a blond guy with a goatee, the only one in the army I saw wearing a beard.

The next morning the first sergeant blew the whistle, got us up, we stood formation and had a head check to see how many men were still available in the company. The captain said we might stay here awhile because he had no orders. At this point the army was regrouping itself, supplies were slow in coming up so no one was moving too fast. The sky had cleared up, it was a nice day, and after breakfast Roper and Riley looked around the factory and decided to crack the safe. It was a Monday but no one had come to work because of us. The *Amis* were here. Roper and Riley were hip-deep in papers, trying to pick the lock, and finally they gave up and put a grenade under it and just blew it up. Inside there was nothing but two albums of stamps! And beside them, just lying there, were a pair of ladies' suede gloves. No money, no *Reichsmarks*, nothing but two albums of stamps and a pair of long, suede gloves that went up to the elbow. Someone had put them away in the safe, and later that day when the Polish women came rampaging through the town I thought of those gloves.

Everywhere the Allies went now in Germany, the people they were liberating were not the local civilians, like the Dutch people in Heerlen, now they were liberating the prisoners, the slave laborers, all the people in labor camps like I'd been in. People were working everywhere throughout Germany, including in this case the stove factory. The ones who worked in the stove factory were Polish women and later that day a bunch of them came parading down the road past the factory. They had been pillaging their way through the town, going crazy, running amok. During the war—for however long—they had lived like animals, wearing rags, hungry and cold, raped and abused, and now they were showing off, parading through the streets in clothes they had taken from the Germans. Instead of rags and wooden shoes, they were wearing all kinds of crazy clothes, and looked so outlandish it was hilarious. One woman went striding by in red boots, the kind of boots that belonged in a stage show or a carnival, and when I saw them I thought of the pair of long suede gloves we found in the safe. Someone had wanted to save the gloves, they had meant something special, a whole way of life someone wanted to protect, and now the Polish women from the camp were strutting through the streets in red boots and stockings and hats.

In the afternoon Ozzie found me and said, "Let's go see if we can find something to drink." He had a three-quarter ton truck and we drove alongside the railroad track into town. "Stop that Kraut and ask him if he knows where there's any booze," he said. I was a mean sonofabitch with a .45 on me and I got out, walked over and hit the Kraut in the mouth with my gun. "Do you know where there's any cognac or wine?" I said. I had hit him just to be ornery, to make him think I might shoot him. He said yes and I pushed him in the front seat of the truck. The poor bastard was bleeding but he directed us to what looked like a store. I rapped on the door with my gun butt. The door opened and it was an old German. "Someone said you have wine and brandy." *Ja, ja,* he said. "Can you give me some coffee and butter?" I said, "Sure," but I told him I wanted to see the stuff first. He took us down to the cellar where he had booze all over the place. He must have been a dealer or had a store somewhere. He helped us load twenty cases of *Branntwein*, a kind of German brandy, on the back of the truck. When we got done, he said, "What about some coffee and butter?" I said, *"Auf Wiedersehen,"* and jumped in the truck. Taking something from a Kraut was a joy and even if I'd had something to trade I wouldn't have given it to him. The way it was before, I had to watch out for the Germans. Now it was my turn. They had to watch out for me.

When we got back to the factory, the whole company came around as soon as they heard we had booze. There was so much the supply sergeant took over. This was a real treat because the army only gave the troops a drink once a week. At the end of the chow line they put out a big pot and each guy got a scoop of whiskey. I'd take a sip and give the rest away but this time when it was my turn I told the supply sergeant to fill me up. "Keep pouring, sarge. Fill it right up." I'd never been drunk but I was going to drink these guys under the table. I chugged it down and five minutes later went back for seconds. Sarge poured a little more in my cup and said, "That's it, Little Joe. I'm not giving you anymore." I was a hotshot seventeen year old kid who thought he couldn't get drunk. I should have listened to the sergeant. Jesus Christ, I got drunk! And sick! The next day I was sweeping up in the factory, and I said, "What the hell stinks around here?" It was my breath. What a foul drunk! I almost killed myself drinking so much *Branntwein.*

We were going to stay in Viersen awhile and whenever we became stationary the quartermaster came around with a big truck to deliver supplies. Jimmy Birdsall and I were on guard duty when a two and a half ton truck pulled up in front of the factory. The driver and the other guy went in to see the mess sergeant to find out how many supplies to leave, and while they were checking the list of guys left in the company, Jimmy and I took the opportunity to jump in and check the truck. It was loaded with eggs, butter, lard—the American army never went hungry—and it also had chocolate. They fed us bitter chocolate to prevent diarrhea, they said, and Jimmy and I each grabbed a five gallon can. We hid them around the corner in a doorway and nobody missed anything. It didn't take long to find something to trade for it. Next door to the factory was a house and because so many German civilians had been bombed out everyone was doubled up and the house was full of families. We saw a girl walking out of the house and she said, "Hi," and we said, "Hi," and we started talking to her, in English which she knew fairly well. Because we were in Germany now and the Germans were the enemy, the army had a non-fraternization policy. I was so naive I believed it—but it meant nothing! Even the war meant nothing! When you're young, you're always in love. A guy sees a broad, he wants to get laid, she wants to get laid—no way a GI is going to pass up a love affair. Besides, the Germans were starving and we had something they needed. While we were talking, the girl asked if we had any chocolate. Jimmy and I looked at each other. Have we got chocolate? She said there were a lot of *Fräuleins* in the house and maybe we could come over and bring some chocolate. We told her we'd be glad to, but we had to wait till dark because of the non-fraternization policy. She went home and Jimmy said, "Shit, Little Joe, we can get the whole company laid!"

That night Jimmy and I went over to the house. We unbuckled the .45s and left them behind so they could see this was on the up-and-up. The girl was there with her friends, they're all happy to see us, and Jimmy says, "Go back and tell the guys there's a party here." I went back to the factory where guys were lying around and said, "Look, fellas, there's a party next door with some frauleins waiting for you." A lot of them jumped up and I went around to where Jimmy and I had hidden the two containers of chocolate. It was a lot of chocolate—we were tipping very heavily—but what the hell, and when I went back, they were making a meal for us, cooking, frying potatoes. These were the enemy but the goddam Americans, how fast they forget, and the guys started running back to the factory to get C and K rations and some of the *Branntwein.* So many guys came to the house the German army could have taken us over. We had a feast, a real party, corned beef, ham and eggs, eating, drinking, screwing. There was a war on, you could look out and see flashes from the guns going off, and here we are kissing and loving them! After a couple hours the sergeant who was the CQ sent over his corporal and said, "You guys better get your ass back because if Captain Hardesty comes around, you're all going to be in jail." Most of the guys went back—some stayed because they had such a good thing going—and I went back too. I crawled in my fart sack, worried I'd get in trouble if the captain found out because about twenty guys were still missing. It was lights out, and blackout, when all of a sudden the door opens and a flashlight shines in, followed by John B. Roper. He's been next door and he's pissed off. "Where's that little sonofabitch!" I heard him and thought, there's only two little guys in here, me and Jimmy, but Big John is a good friend of mine so he must be pissed off at Birdsall. "Where's that fucking Kraut-lover?" Someone said back to him, "Come on, John B., shut the goddam light off." Although he didn't say it too forcefully, because John was pissed off, and he was smashed too. "I'm not going to bed till I find that fucking Kraut-loving Little fucking Joe!" I said, "Shit!" and crawled down to the bottom of my fart sack and prayed he wouldn't find me. Because he was serious. We had given the frauleins chocolate and he wouldn't give a German spit. He had seen too many of his buddies get killed, so his attitude was, you want to get laid, you fuck her, you rape her. You don't love 'em and give 'em candy. John B. stumbled around looking for me for a few minutes before he gave up and found his own fart sack. No one got caught and Captain Hardesty never showed up. He was probably doing a little fraternizing of his own.

Finding frauleins and booze was something I could do for the guys in the company and I was always trying to do things to get close to them. At the time I didn't think about it—I didn't think, I just acted—but looking back I can see that sometimes shrinks have the right interpretation—it was my desire for love. It was like when I was chasing Basha at the age of thirteen. I just wanted that kind of contact. We were fighting a war and the guys were rough combat soldiers but it was the first time I'd been treated decently in my life. At home when my father took his rage out on me, it was him and me. Now it was totally different. I had a gun in my hand and two hundred guys backing me up. I was taken care of, and I was treated like everyone else. I got my mess kit, stood in line, and if I happened to be first that was okay, and if I was last it was my own fault. But I was no different, I belonged where everyone else belonged. I was one of the guys in the unit and if there was anything I could do for them, I would do it.

By early April when we got to a small town called Selm further to the east, a good hundred kilometers, the German army was retreating and blowing everything up behind them. We had crossed the Rhine on a pontoon bridge and had some serious fighting on the way from Viersen but by now the Russians had taken Berlin, or were knocking on the door, and most of the time we were picking up German soldiers who were surrendering. We disarmed them and sent them back to prison camps. By now the Germans were so desperate to find troops that some of their soldiers were as young as me. But they were just kids, they hadn't been through anything, and when we caught them they were so scared they'd be crying. We cut off their pants at the knees—they were too young to be wearing long pants!—and told them to go home.

In Selm we moved into a hotel where we were served by about thirty Polish women, what they were calling "DP's", displaced persons. They were the slave laborers who'd been liberated by the Allies. The Polish girls were beautiful and the guys' tongues were hanging down to their navels but the guy in the outfit whose family had come from Poland got very protective and he wouldn't let anyone touch the girls. "If you want some Kraut pussy, that's okay, but don't screw around with my people," he warned everyone, and he meant it. All these Polish bunnies were walking around, waiting on us, making our beds, and no one could touch them. I think they felt they had a little too much protection, too, and wouldn't have minded some fraternizing.

The Polish women were part of the disarray in the whole area. Besides the German soldiers deserting, there were German civilians bombed out of their homes, and especially there were the hundreds and thousands of DP's liberated from camps. It was chaos because whenever the Allies took an area there would suddenly be hundreds of these people liberated from labor camps. Just like in Viersen they would go wild through a town, take things at random from the Germans, go into homes and confiscate whatever they wanted, and beat the hell out of anyone they found. I knew their anger because to the DP's the German civilians were like the ones who had jeered at us on the way to the coal mine when I was a prisoner at Kohlscheid. Now the prisoners were suddenly free but they still had nothing and it led to many mob scenes. While we were in Selm we were called to quell a riot of DP's. We drove to the location in jeeps and found at least a hundred DP's dragging things out of houses and loading up wagons and pushcarts to take back to where they were billeted. It was an ugly scene, beaten-up Germans and stuff lying all over the street. We were ordered to form a line and walk up the street. We fired our rifles in the air and told them, "Go back! No more looting!" They left but we let them keep whatever they'd taken. The next day the Germans started coming to us complaining and Captain Hardesty had me translate for him. They said they had been beaten up, their radio or bicycle had been stolen. Hardesty told me to tell them they were lucky to be alive, and they got the message. They were still afraid of us.

By now I was established in the role of translator for Captain Hardesty. He had orders to pick up anyone who looked like a soldier and he was such a hotshot he was always bringing guys in and having me interrogate them in German. Yes, they were in the army, yes, they were hurt or deserting. So the MP's came and picked them up. Then the MP's drove around the corner and kicked the Krauts out of the jeep. It was too much trouble to detain them, so they just let them go.

Another order we were supposed to enforce was that the Germans had to get rid of *Mein Kampf*. No one was actively looking for books and many Germans may have burned their copies before we got there, but I was standing guard in front of the headquarters in Selm when a German civilian came up and said, "I'd like to speak to the commander."

"What do you want?" I said.

"I know of someone who has a copy of *Mein Kampf* in his house," he said.

I sort of wanted a copy myself, so I went in and said to Sergeant Hannah, "There's a guy outside whose neighbor has a copy of *Mein Kampf*."

"My what?" Hannah said.

"You know, that fucking book Hitler wrote."

"So what?" Hannah said.

"They're not supposed to have it," I told him.

"They're not supposed to have it," Hannah said back to me. "Well, go get it, for chrissake!" Not that he gave a shit about some book.

I put the guy in a jeep and he directed me where to go. I knew how to drive by now and like all the GI's I drove like hell, up on two wheels around turns, scaring the hell out of this Kraut who probably thought he'd turn someone in and make points with the *Amis*. He showed me the house, I knocked on the door, and a guy about twenty-eight or -nine answered. I pulled out my .45, scared the shit out of him, and said, "You've got Hitler's *Mein Kampf*."

"How did you find out?"

"Just give it to me," I said.

"I was going to destroy it," he said. He went in the other room and came back with the book, beside himself with fear.

It was a nice, cloth-covered copy and inside the cover it said, "To *Gefreiter Schweicker*, on the second war Christmas."

"Are you Schweicker?" I asked him.

"Yes."

"What'd you do to get this book?"

"I didn't do anything."

"You must have done something to get it."

The poor guy. I was being a real punk.

"No, no. Nothing. The company commander gave it to everybody."

I took it and left.

I didn't begin reading *Mein Kampf* until after the war was over and we were in Wetzlar. I was too young and ignorant of the history and politics to understand the background of events Hitler referred to, like the Versailles Treaty at the end of World War 1, so a lot of it didn't mean anything to me. But what I could understand was Hitler's rabid hatred for the Jews and his philosophy about how to use the Slavic people as a labor force. The Germans would be resettled to the east, in Poland and Russia, and in the new order the Slavic laborers would work under German administrators. All this was spelled out very clearly. And that was something I could understand, obviously, because that was what I had been doing in the coal mine in Kohlscheid, working as a slave for the German *Reich*. So what I understood were the racial parts of his program, and also the fact that he had written this in 1925, long before the war. Then he had done just what he said he would do. I couldn't understand why leaders and nations hadn't reacted to Hitler, because he had spelled it out in this book, and millions of copies had been published. He described how the Jews were the most unmotivated, unartistic people, that they couldn't do anything from within themselves, that whatever they did in everything from music to art they had to borrow from other societies because they no culture of their own. He was such an incredible fool—yet this book was used as a textbook in school and when you got married in Germany you got a copy as a gift from the *Führer*. I didn't talk to anyone about it except Major Gingold who saw me reading it one day and asked me how far I'd gotten, but we didn't discuss it. Perhaps it made him too upset. He was the only one who showed any interest. The evaluation of the whole Nazi program for guys like John B. Roper or Riley was just, "Those fucking Krauts..." If the sonofabitch wore a German uniform he was the enemy. That's as far as it went. But for me, reading *Mein Kampf* showed the thinking or philosophy—if you want to call it that—behind some of the things I had seen and how I had lived.

After a week in Selm we moved out for Bielefeld. It was a large town another hundred kilometers east and when we got there we camped in a former German army barracks. The Americans and Russians had already met at the Elbe River and early in May, on a nice sunny day, we got the news we were waiting for. I was watching some GI's playing baseball on the parade grounds in front of the barracks when all of a sudden I heard machine gun fire, pistol fire, and M-1s all firing from the British unit bivouacked beside ours. Guys were screaming and shooting and we all grabbed for our guns, thinking we were being attacked. The Brits were waving and firing in the air, but it still didn't dawn on us what was happening. Then Sergeant Hannah came running out and said, “The fucking war is over!” Everyone went apeshit and I put my M-1 on automatic and PSCHEW! I shot off a clip. The captain ran out and shouted, “Filipowic! What the fuck are you doing?” I was doing the same thing everyone else was doing, only I was doing it right in front of his office. I yelled back, “The fucking war is over!”

The war was over! While everyone was celebrating, Ozzie pulled up in front of the barracks with a 3/4 ton truck loaded with wine. The guys broke out the bottles and some of them got German army uniforms from the barracks and ran around shouting, “Heil Hitler! Heil Hitler!” and giving each other the Nazi salute. All the units went crazy and celebrated into the night. Everyone got smashed—except me. I’d done such a job on myself at the stove factory in Viersen I hadn’t touched a drink since.

The date was May 8, 1945, five years almost to the day since the Germans invaded Holland and I woke up at dawn to run out and see the plane flying over the house.

Chapter 8
American Zone, Germany; May 1945

After V-E Day, the Allies divided Germany into four zones, American, British, Russian, and French. Bielefeld was in the British zone so we got orders near the end of May to move out, first to Königsburg, then to Wetzlar in the American zone. Our column made its way through hordes and hordes of people on the roads. Even the back roads were full of people, and the movement continued for months. There were thousands of DP's liberated from labor camps, both men and women, and there were German families bombed out of their homes, often with children, walking and pushing bicycles loaded down with the belongings they had taken with them when they fled during the fighting. Now they were trying to make their way home again. We also saw many German soldiers still in uniform, disarmed of course, walking along the roads and hitchhiking home. Instead of putting them in makeshift prison camps, the Allies had just processed and released them. Many were older and looked beat; they were totally defeated men. There were huge German diesel trucks on the road, pulling two or three trailers hooked together and loaded with people. We also saw Russian convoys, one after another, hauling thousands and thousands of Russians back home. Only they never went home. Stalin sent them all to camps in Siberia because in his mind the POW's and the DP's had all been contaminated by contact with the west. They had a very bad fate.

A lot of the Germans were worried because they didn't know what would happen now, and what they heard on the radio or by rumor wasn't encouraging. The talk was about the concentration camps, of the atrocities that had been committed, and of course all the forced laborers were letting out their anger on the Germans. The same rioting we'd seen in Selm was happening many places, although where before the DP's would go on a rampage in mobs, now it was more common for five or six guys to go out and steal something, grab someone, kill them. The Germans themselves very much wanted to get out of the Russian zone because they had heard stories about the Russians raping them and pushing them around. What was happening there was not nice. The order of preference for the Germans was first the American zone, then the British, then the French. The Americans were lazy and they had food, and if you worked for them doing KP or greasing trucks you always got a meal. The British soldiers didn't have all the food and cigarettes to pass out or trade, and the French had even less. Rationing hadn't been set up yet, stores had nothing, there was no commerce, no transportation—how would you get produce into a city? No way. So that was everyone's main concern, food. Food and shelter. It was nothing to go into someone's home that looked decent, to stand in the living room, open a door and you were looking outside. Half the building was gone. One step and you'd wind up in a pile of rubble.

We made our way through the chaos to Wetzlar. It was an old town on a narrow river called the Lahn and the only industry of any consequence was the Leitz factory where they made Leica cameras, so it hadn't been bombed at all. Word had come down before we shipped there that Captain Hardesty had really found us a place. "You won't believe it!" they said. When we got to Wetzlar we drove through the town, left it behind us, followed the river, then took a dirt road up a high hill toward a steeple. On top of the hill we drove past several houses and a farm, then in through a gate to a square formed by low buildings attached to a church. It was an old medieval convent, a beautiful place overlooking the Lahn River valley, with hills rising up behind us.

The Germans had used the convent as a camp for forced labor and when the Americans took over they put up a new three-room concrete building on one side of the square. Sergeant Hannah promptly moved in and slept in the living room, I had the bedroom, and there was a bathroom. Now we were living! Every GI had a room of his own, there was a dining room the nuns had used, and a kitchen for the cooks. Within a week we were set up and going. We had Germans working for us and there was no duty of any kind. We stood reveille every morning and that was it. Wait for the pork chops at lunchtime and see what's for dinner. We organized a service club in the day room and although the rule was still no fraternizing with the Germans, it was just like before, everyone ignored it. I was in charge of getting frauleins. That meant driving down to the town with Chief, the Indian in our outfit, and whenever we saw a girl we'd stop the truck and I'd ask her, "Do you want to go to a dance?" We'd pile the broads in the truck and drive back up the hill. For a dance we needed music, so when we saw a young guy in Wetzlar I stopped him and asked if he knew anyone with a band. He said he had a friend about ten miles from Wetzlar with a band, so I told him to show us where he lived. He wanted to get in with the Americans, so he jumped in and while we rode along he asked if I was German.

"No, I'm not German," I said.

"Were your parents German?"

An American GI was speaking fluent German to him and he couldn't figure it out.

"No," I said.

Chief asked me what the guy said.

"He thinks I'm a Kraut."

Chief thought that was very funny.

The kid directed us to a nice home in a suburban village where a middle-aged woman was raking the garden in her front yard. He got out and asked if Heinz was home.

"Ja. Was ist denn los?" she said. What's wrong?

"Nothing, but these Americans need a band every Saturday and maybe Heinz wants to play."

"Play for the *Amis* and take his drums? You're crazy!"

"Be careful," the kid said. "He speaks German."

"How well do you speak German?" she asked me.

"Just go get him," I told her. "We're going to take him with us."

By now Heinz had come out and his mother said, "Don't go, they're going to take your drums!"

I said, "Go get your drums. Don't worry, we won't take them—we want you to play."

He piled his drums in the back of the truck with the other kid, and Heinz sat up front and showed us where the other band members lived. We found every guy, six of them, and while we were picking them up, Heinz said, "We also have a singer."

"Yeah? Where's the singer?"

We picked her up too, and she was a knockout. I said to myself, if we bring this broad back with us there's going to be trouble.

We drove back to the convent and I told the band if you play for us you'll get something to eat, get paid in scrip, and we'll give you a ride home at night. It sounded good to them. It was good for us too because by now all the GI's were setting up clubs and there was a mad scramble for musicians. While we were making arrangements, Captain Hardesty walked into the day room and saw the singer. "Filipowic," he said, "who's the Lily Marlene?"

"She's the singer."

"Ask her if she wants to fuck tonight."

I went over and said, "The captain wants to sleep with you tonight."

She gave me a look and didn't answer.

"Tell me yes or no," I said.

"I can't say that I want to do that," she said.

"Okay. But don't get involved with anyone else, then."

She looked at me and then sort of gave Captain Hardesty a little smile.

I went back over and told the captain I thought it was going to be okay. "But," I said, "I promised we'd take her and the band home tonight because we want them to come back next Saturday."

"Don't worry," the captain said. "I'll work it out."

Which he did. He laid her that night, I drove her and the band home, and two weeks later she was living with him. She became a standard fixture—she never got out of his bedroom! The captain had breakfast at the officers' table and when he got finished the cook gave him another tray and he carried it into his room. Same thing at lunch. And supper. Hannah even yelled at him once, "For chrissake, let her see some daylight once in a while instead of staring at the ceiling all the time!" Within a span of two months the captain had two broads living with him, the singer and another one, both real good looking. The warrant officer had a woman living with him. The lieutenant had a woman. It seemed like everyone had a woman in his room. At meals guys would eat, then go back for seconds and take the tray to their room! I didn't have to go out as much on Saturdays looking for frauleins because they were all living at the convent. The band strikes up and they're all right there dancing!

My biggest job was finding booze. You can't have a club without liquor and the army was not about to supply us with that. All the other units were looking for the same thing because we didn't even get beer, only Coke. So we'd load up the truck with flour, sugar, coffee, chocolate, and lard, and drive to the French zone. Chief always drove and I had regular vineyards where I'd trade food for wine. The owner would give us a little taste of each barrel, so by the time we drove back we were already smashed from tasting the wine. We always had a couple bottles up front to drink on the ride back too. Early one Saturday we left the convent, drove to the French zone, made our stop at the vineyard, loaded up, and on the way back we saw two girls walking along wearing summery dresses. Chief stopped and I asked them if they wanted to come to a party. We were quite a ways from Wetzlar but they said yes and climbed in. We were all jammed together up front in the open truck drinking wine, one girl in my lap, and above us the vineyards covering the hills.

Along the edge of the road was a railing and below us was the Rhine River. Coming toward us I saw one of the big diesel trucks pulling two trailers, and Chief laughed and shouted, "Look at that big sonofabitch!" I shouted back at him to look out because the railing was getting pretty close, and the next thing I saw was our right front wheel rolling down the highway up ahead. I didn't even feel us hit the railing and by then I was flying through the air. I landed in the river. The water was low and the truck landed on its side. One girl was moaning, the other was up on the road screaming. I was sore all over but nothing was broken. Chief was lying under the truck and I could hear him moaning too. I got up and the way the truck was balanced when I gave it a shove, sure enough, it tipped over. The German truck had stopped up above on the road and the driver and his helper came sliding down the embankment. Chief couldn't move, even with the truck off him. The one girl had a couple gashes in her leg and the other was still up on the road screaming. Wine bottles were lying all over the place, the water lapping at them as boats went by making waves. The embankment was so steep—almost like a wall—we couldn't carry the girl up to the road so the two Germans got a rope, put it around her back and legs, and pulled her up. Somebody else had stopped up on the road and they drove the girl and her friend to a hospital. Then the Germans tied the rope around Chief's ass and under his arms and pulled while he sort of crawled. By now people were all over the road. They'd come up from the village when they heard an American truck was in the river. A woman came over and said in English, "Can I help you?"

I said, "I want to take this guy to the hospital." Chief was lying by the side of the road.

"I'm a nurse," she said. "He doesn't need to go to the hospital. Why don't you just take him to my house."

She lived nearby so we got Chief to her house and laid him down on a bed. She took off his pants and started washing his legs which were pretty well scratched up and mashed, and Chief lay there with a big grin on his face.

"I've got to get back and tell Hannah what happened and see if we can get the truck out of the river," I said to Chief.

"Okay, you go back," he said.

He wasn't going anyplace. He was still smashed and didn't know what the hell had happened.

I hitchhiked back to Wetzlar and when I got there Hannah said:

"Where's the truck? What happened?"

I told him the story. The wheel came off; we went in the river.

"Stupid sonofabitch!" Hannah said. "Get another truck. And get Chief!"

I got another truck and Hannah and I drove back. Because the bank was so steep there were still a few cases of wine left down by the water, although some of the local Germans were trying to retrieve them. We chased them away and hauled the cases up with a rope.

"What about the truck?" I said.

"The hell with it, leave it there," Hannah said. "Let's go get Chief."

We went to the German nurse's house to pick him up but Chief still wasn't moving. "I want to stay here."

"Get your ass back," Hannah said. "The captain wants to see you."

"Fuck the captain. I want to stay. I've got it made here."

"You may have it made but if you don't get back you'll be AWOL."

"The hell with it, so I'm AWOL."

We finally persuaded Chief to come with us—but only because the woman came with him. She was about forty, nice-looking, and whatever Chief did for her, she wanted more. She stayed at the convent and lived with him for a month or so.

Chief was a good guy but a bad drunk and those were the times Captain Hardesty had to be aware of him. When Chief got drunk he'd come looking for the captain to challenge him. Hardesty may have been heavyweight champ of the Hawaiian Islands before the war but when Chief got drunk he always wanted a piece of him. And the captain was smart enough to avoid him. One Saturday night Chief got drunk, took the jeep with the .30 caliber machine gun mounted on the back and drove into Niederbiel just outside Wetzlar and shot up the village. We got a frantic call, "There's an American running up and down the street shooting everything!" Captain Hardesty and Sergeant Hannah and I looked for the jeep—sure enough it was gone—so we got our rifles and raced down the hill in the other jeep to Niederbiel. On the way we could hear a machine gun rattling. We stopped a couple blocks from where the noise was coming from, because we didn't know who it was, and started shooting up in the air. The machine gun stopped and the captain yelled, "This is Captain Hardesty—what the fuck are you doing out there?"

A voice yelled back, "I want to get myself some Krauts!"

Hannah said, "Oh, shit, it's Chief!"

Hardesty shouted back, "Stop before I shoot your ass off!"

Hannah drove up and got him. I drove the stolen jeep back and they put Chief in the other one with the captain in back holding his rifle against the back of Chief's head. If he makes a move, the captain will blow his brains out. Because we didn't know if he'd shot anyone or not. Hannah put him in the orderly room and the captain told the charge-of-quarters not to let the sonofabitch out. "If he goes to the bathroom, go with him."

We drove back to Niederbiel and people were all over the street, shouting back and forth to see if anyone was hurt, and complaining about the damage. Windows were shot out and busted, there were holes in the walls, bullets had gone through china closets, the street had been strafed. But no one was hurt. The captain gave Chief "company punishment" which meant he couldn't go off the post for a month.

Whenever there was trouble, they always took me along in case they needed a translator. Sometimes a few GI's went down to the village and cleaned out a beer hall, or they'd wait by the movie theater and when the show got out they'd pick on the ones who looked like they'd been in the army. But most of the calls we answered were from Germans frantically reporting DP's who'd gone apeshit and were beating or stealing. One day a German rode his bicycle up the hill and said, "Come quick, the Polacks killed a man and stole his pigs." We put his bike in the back of the 3/4 ton truck and raced to a farm in Niederbiel. Sure enough a man had been stabbed to death and everyone was in an uproar. They wanted us to go looking for them but we didn't bother. The captain wasn't going to go looking for DP's. If he found someone, what was he going to do with him? As long as they were only using knives he wasn't concerned. If they used a rifle or gun he had to follow up because no one was allowed to have firearms of any kind, no matter who they were. But if someone stole a pig and a German got stabbed that was just tough shit for them, and if anything, the captain felt the DP's were right, after what they'd been through.

Germany was still in chaos. They were trying to sort out the DP's so they could be repatriated, and while they were in the camps they got a food ration, but it wasn't much so it was common for DP's to go out on raiding parties, kill a cow, skin it, cut it up, and take it back to the camp. Things were tough for the Germans too and the mood among them was very grim. Their food ration wasn't enough to live on and they had no money to buy anything. The American zone in western Germany had little farmland. The saying was that the Russians got the farms, the British got the industry, the French got the wine, and the Americans got the scenery. So the best thing for the Germans was to work for the Americans because then they could eat. The guys in the band and the frauleins we picked up were glad for a chance to get close to the Americans because that meant food. The army brass didn't want GI's pulling KP duty anymore so once we were situated at the convent we were entitled to German POW's to work in the kitchen, in the motor pool fixing trucks, as orderlies for the captain, and they also worked at the service club as waiters and bartenders and did the cleaning up. We had thirty-five former German soldiers wearing dyed fatigues with "POW" printed on their backs and because I could speak German I was put in charge of them. I was also one of the permanent guards, although we only bothered to guard them at night, in shifts. They were quartered in a wing of the convent that abutted the church, on the second floor, and they weren't mistreated at all. They were hard working, they got fed, we gave them cigarettes, so they lived pretty well. They had been working for us for a month or so when one night Captain Hardesty came by and said, "Little Joe, open the door, I want to see what those Krauts are doing."

I opened the door and all the Krauts were gone! Only about six were left and they were white, scared to death.

"Ask them where the others went," the captain said.

"Where're your comrades?" I asked.

They said they didn't know.

There was a stairway down to the church sanctuary below and the captain went over and lifted up the big boards we'd used to barricade it.

"You stay here, Filipowic," he said, and went to get Hannah and Riley. They drove a jeep up to the church, opened the door and shined the headlights in—and there were all the POW's making out with frauleins in the pews! The girls had come up from the villages and through the woods to the church. We made the POW's stand formation, had a head count, and they were all there. After that we put a guard outside the church. But no one really cared if they escaped—and why would they try? They had it made, working for the Americans, and the only reason any would leave would be to go see their families.

The other job I had—if you can call it a job—was going back to Holland with Sergeant Hannah to trade cigarettes for booze. His buddy the master sergeant who worked for the quartermaster was supplying him with the cases of chocolate and bags of sugar and coffee and lard that Chief and I traded at the wineries, and he also got a few extra cases of cigarettes and C- and K-rations every week. To peddle cigarettes in Holland we had to drive through the British zone but that was no problem because all you needed was a trip-ticket and Hannah always had a stack of them in his pocket. They weren't legitimate because he'd just fill them out and sign them himself and we'd take off. Sometimes we had so many cases of cigarettes we pulled them on a two-wheeled trailer behind the jeep. It was about four by six feet with a tarpaulin tied over the the top. We took the back roads to avoid the MP's who would impound the cargo right away if they inspected it and get us in serious trouble for stealing government property. At the border the Dutch army soldier on duty often just waved us through.

We had made a couple trips back to Holland before the end of the war and they became more frequent when we were at

Wetzlar. We sometimes went different places, a couple times as far as Liege, Belgium, where there was night life and gambling and GI's all over the place. On the way if we saw some frauleins riding their bicycles, the sight of those beautiful limbs would make Hannah's animal instincts get the best of him. He'd stop the jeep and I'd speak to the frauleins, tell them we've got cigarettes and C-rations, and they were always very happy to accommodate. We screwed our way across Germany. Under normal circumstances, I'm sure the way a guy like Sergeant Hannah grew up, or me too, you couldn't just stop a girl on the street and get laid. But the Germans were starving, they were desperate, and for soldiers like Hannah, in combat from Africa to Sicily to Normandy to Germany, their buddies had always been dying and they thought the next piece of ass is going to be the last one. It didn't matter if the guy was a truck driver or a Princeton graduate, the one thing he wants before he goes is a woman. And for the German women, their men were constantly going and never coming back, so before he leaves what can she give him? People did things that under normal circumstances they never would have thought of doing. What was taboo a few years earlier was not taboo anymore. The bombers could come over at night and wipe everything out. One bomb and you were gone. So there was no vulgarity or ugliness in it, the way it was portrayed before the war. It was romantic in a way, a human romanticism. People were just reaching for something. It even made me think somewhat differently of the collaborators. My attitude before had been, "Those fucking broads are screwing the Germans." But maybe they had deeper thoughts than I did. Maybe the women were more sensitive to it than men. Young people were dying all the time, there was nothing else, and that was the one thing you had going for yourself. After the war it was somewhat different but the way Americans were treated in Holland and Belgium was incredible. For whatever reason, the GI's were so generous with women, with children. They put the food out there first, and if they got some affection, fine. Just like on our trips through Germany to Holland. If we saw a good looking woman on a bicycle, we didn't knock her over the head like a Neanderthal, we stopped and talked. We didn't threaten them, we weren't ugly about it, and they were glad we'd stopped. Afterwards they would say, "Do you make this trip often? Next time, please—here's our address, please stop and stay with us."

The first time Sergeant Hannah and I went back to Heerlen, we located my friend Vwadzu to find out who was buying black market cigarettes. Vwadzu couldn't believe it when he saw me in an American army uniform with a .45 on my hip, a steel helmet, riding around in a jeep with Sergeant Hannah. The last time he'd seen me was on the day the *Amis* arrived to liberate Heerlen, at the party at Leo's. Vwadzu jumped in the jeep with us, asking me questions about the *Amis* and what I was doing while he directed us to a very fashionable home in the center of Heerlen. The guy knew Vwadzu and although he was surprised to see these two "Americans" he figured we weren't there to arrest blackmarketeers, so he invited us in. He gave us a drink of *Schiedam* Bols gin, top shelf stuff that Hannah loved—he'd developed a taste for Dutch gin—and we made a deal. We didn't want to be driving around with the cigarettes any more than necessary, so we unloaded the trailer right there in broad daylight—cases of Chesterfields, Camels, Old Golds—and he gave us a big roll of *guldens*. It was all the money he had, but it wasn't even half the price because we had eight cases of cigarettes, a hundred cartons to a case, and he wanted them all. If we come back tomorrow he'll have the rest of the money. I said to Vwadzu, "You know the guy, is he okay?"

"He's okay, Josko. If there's a problem, you can always come back and blow his brains out."

Vwadzu was joking but the effect of the uniform, the Jeep, the gun, it all made him think Sergeant Hannah and I could get away with anything. But we could get in serious trouble selling cigarettes on the blackmarket. The trip-ticket was phony—Hannah typed up the traveling orders and signed them himself—no way we belonged there.

We drove Vwadzu back to Frankstraat. I had no intention of going to the house at #22 but he told me the news. Dragitsa was working in a hat shop in Heerlen and had a baby, and my father had been in jail because after he found out she was pregnant he stabbed her Dutch boyfriend with a knife. Fortunately, the guy lived but my father had to serve time. They put him in jail. Dragitsa had married the Dutch guy but she was still living at home because her husband was in Den Haag working. I knew where the hat shop was but I wasn't going to go see her. I didn't ask Vwadzu about Gizela, if she had survived. She was always there in my mind but I hadn't come back to see anyone at #22.

"How is Basha?" She was Vwadzu's cousin and I hadn't seen her since the night at Leo's.

She was okay, Vwadzu said. She and her parents had made it through the war. "Why don't you stop and see her?"

"Tell Basha I'll come over tonight," I told Vwadzu when we dropped him off.

Sergeant Hannah wanted to find a hotel where GI's could stay and find some action so we checked into a nice, plush hotel in the center of Heerlen. All this was incredible for me, because the hotel was the kind of place where before the war, if Josko had dared to step in the lobby they would have thrown him out in the street. But when I walked in with Sergeant Hannah, we were treated like royalty. "Yes, sir, can we help you?" "We'd like two rooms. Nice ones." And we had the money. Hannah's got a roll so big he can't get it all in one pocket! And I've got another one! Instead of throwing me out, some guy wants to carry my bag. I make sure he gets a big tip.

After we checked in, we went to the Holiday Restaurant. It was all GI's. The Dutch didn't even come in anymore. Hannah and I had a drink and I looked around and saw a waitress I knew. When she went by I said, "Hello, dear."

She looked at this guy in the American uniform and said, "I don't know you."

I said, "Yes you do." It had been awhile and she was a couple years older than me, Dragitsa's age, but then she recognized me and it was just like with Vwadzu, she couldn't believe it.

"Josko! Is that really you?"

Right away she had questions, what happened, what am I doing in the American army?

"I've got a friend who wants to meet you," I said.

I introduced her to Hannah and they arranged to meet later when she gets off work.

After a couple of drinks, Hannah said, "I want some champagne, Little Joe."

"Okay, let's find out who's got it."

I asked a few questions and got the address of the blackmarketeer who had champagne. We found him, bought a few bottles with the cigarette money, Hannah went back to the hotel to kill some time before he picked up the waitress, and I went to see Basha.

When I left #22 to hide out in the water tower I told Basha I wouldn't come back till I could walk up the street like a man, and that's exactly how I felt walking up Frankstraat wearing the American army uniform—I wasn't just working for the Americans or getting close to them, I was *with* them, I was *one* of them. I trotted up the street, half-crocked, with two bottles of champagne, shaking them like crazy—a real bigshot!—ready to show Basha and her parents how the other side of the world lives. I'm wearing my OD's, the formal army uniform, Vwadzu has told them I'm coming, and they're so glad to see me—the table is beautifully set with the best crocheted tablecloth, handmade in Krakow or wherever, the best dishes are out, her mother is cooking and Basha is smiling and her father is shaking my hand. Josko sets the champagne on the table, takes off the wire, and no sooner is the wire off than the bubbles press on the cork and PSHHHH! half the bottle of champagne ends up on the ceiling! It blasted the whole room, raining down on the tablecloth. It was the easiest cork that ever came out of a bottle of champagne—and I really showed them how sophisticated I was!

The people I saw when I came back with Sergeant Hannah wished me well because they knew how I'd grown up and they were pleased to see what had happened to me. One of the first people I went to see was the teacher, *Mijnheer* Van Dijk, who'd gotten married and had an apartment in Heerlen. He was all right, the Germans had left him alone, and like everyone else I met he was surprised I was still alive and even more surprised I was in the American army uniform. "What are you up to now,

Filipowic?" he said, with a big smile on his face. I also saw Leo's brother Scheng who was my age, and Vwadzu's brothers who I had seen the night the *Amis* liberated Heerlen. Taduscz worked for a butcher and Januscz, the smart one, had gone on to high school. Leo was the only one who had been out of Heerlen during the war. None of the others got called by the *Deutsches Haus* or picked up on the street. I also heard the bad stories. I learned that Tutschki had been the first German soldier killed in the battle in Holland when the Germans invaded. His father also was killed in the Dutch campaign. As much as everyone hated the Germans, they were saddened to hear about Tutschki. It made me think of the saying, "Only the good die young." Had his father not lost his job in the coal mines and been sent back, neither of them would have been killed, and everyone liked to think that Tutschki wouldn't have been the kind of German to turn against us.

The saddest one among my friends was Jan te Poel. I had already heard he joined the Dutch *SS* which was a great surprise and disappointment to me because we had been so close. He was like my shadow; I was the one who walked him home from Boy Scouts and school and protected him from the tough coal miners' sons. When the Germans invaded he was very much against them but for some reason he changed. When I saw Jan after the war I was in the American army uniform and he was standing in Frankstraat talking to Kurt, a German guy whose father had been killed in the coal mine. Kurt had never joined the Hitler Youth but he and Jan were good friends. It was embarrassing when Jan saw me but I crossed the street to talk to him. He was standing by the wall of Kurt's house with his arm in a sling and one eye was missing. He'd been shot up on the Russian front and had a glass eye. It was an awkward moment for him too, and there was nothing of triumph in it for me to see him in such a condition. What I felt was a great sadness. It was sad not only because we had been so close and then he joined the *SS* and got badly hurt, but for some reason his whole family had fallen apart. They were the "nice" family, the only respectable one on the street, and Jan and his sister were the ones who would make it. His sister and Gizela and the schoolmaster's daughter were the three brightest ones in the school. But for some reason after the Americans came, Jan's mother went berserk and laid anything that came along. I even saw her once just after I got picked up by the *Amis* and we were still bivouacked at the schoolhouse in Heerlen. I was walking across the railroad tracks when Mrs. te Poel came riding up behind me on her bicycle and when she saw me she said, "Josko, do you want a cigarette?" I told her no because I was issued seven packs a week by the *Amis*. She said, "Stay here, I'll get you one." She left her bike and went across the field to where some soldiers were dug in, the black supply troops. She was gone for a few minutes and when she came back she had two cigarettes and she gave me one. It was so unreal, the tragedy of this—this was Jan's mother, someone I respected so much, from a family so superior to me—and to think this could happen to the most prominent people made me so sad. She had gone from being an upstanding lady to doing this—and she gave me a cigarette I didn't need. So when I saw Jan I was confronted again with the way his family had fallen apart. Not only because he had lost an arm and an eye and was shot full of holes, but because the Dutch had even less use for the collaborators than for the Germans. The Dutch didn't do anything to him because he was such a physical wreck, but his own people would never forget what he had done and there was no way for him to escape his past. His life was ruined.

Sergeant Hannah and I began to make almost regular trips to Heerlen while we were living in the convent at Wetzlar. It wasn't that close but I knew the place so we could always operate on the blackmarket and have a good time. We'd roll in at two in the morning and go wake up the owner of the biggest liquor store to buy the Bols gin Hannah loved. The owner would double the blackmarket price, but we always had plenty of money. When we sold cigarettes we couldn't change the Dutch money into anything so we'd have a wild time eating and drinking our *guldens*. The other thing I did with the money was give it to my sister. When I went to see Dragitsa at the shop where she made hats, I gave her a big roll of *guldens* and treated her to a night on the town, eating and dancing. Like everyone else I saw, she was impressed with the uniform and how I had changed. I hadn't seen anyone in the family since before I got picked up and taken to the camp in Kohlscheid. I never thought about any of them, only Gizela, and all the time I was away in the camp and then fighting the Germans I had lived with the hope that she would get well. When Dragitsa told me she had a daughter named for Gizela, I learned what happened. She never got well, she lay in bed the whole time and finally died at the end of the war, just after the Germans surrendered, when I was in Bielefeld. The family had no money to pay for a funeral so the Pieters who owned the store in Heerlerheide paid for it. Dragitsa said it was the biggest funeral they ever had at St. Anthony's. Besides everyone from Frankstraat and Huiskens, people came from Heerlerheide and Heerlen to show their respect for Gizela, and what it meant that she had died. It was a relief to know she wasn't suffering any more but the feeling I had that I should have done something for her, that what happened to her was my fault, I lived with that all the time.

After I saw Dragitsa at the hat shop, I began stopping by on our cigarette runs to give her some money. There was a friendship between Sergeant Hannah and me by now and he often came with me to the shop. Dragitsa had picked up some English and we would take her out to eat, and as she got to know him she told him about Gizela and the family, so he began to understand some of my anger. He also heard her bugging me to come home, telling me I should stop at the house. She was still staying there while her husband worked in Den Haag, and Mama took care of her daughter. I kept putting Dragitsa off, and if it was up to me I don't think I ever would have gone back to #22, but on one of our trips when we drove into the outskirts of Heerlen, Hannah said, "Where's the house, Little Joe? I'm taking you home." It was two in the morning by that time—we'd just thrown a couple cases of cigarettes and food in the trailer and driven straight from Germany—and when he drove up Frankstraat and parked in front of the house, I picked up a handful of gravel from the street and threw it at the upstairs window. My aunt got up, opened the window and said, "Who is it?"

"Josko."

"Josko? It's Josko!"

She came running down in her nightgown and opened the door. Dragitsa was there with little Gizela, they got up too, and when my father came down instead of Josko he saw an American GI with a .45 on his hip standing in the doorway. Out in the street was a Jeep with a huge soldier leaning against it.

I had no idea how he would react when he saw me. I had no idea how I would react either. The last time I'd seen him was more than three years ago, and that was the night Wil and Leo pulled me off him and dragged me out in the street.

When my father came into the front hall, he didn't say anything. He just walked up and put his arms around me. He hugged me and kissed me. Never in his life had he done that. I hugged him back, and kissed him too. At that moment there was suddenly a change in him, and inside of me too. I wasn't afraid of him, and he was so proud, he was like a peacock. For the first time I felt that I had a father, a real father. It was a feeling—it was something I had never had, something I had been looking for for years.

A party began that went on through the night till the next day. Hannah and I drove into town to our contact to trade our cases of cigarettes, and after we got our *guldens* we bought booze—champagne and liquor—and came back to #22. We brought in the food from the trailer, cans of meat and hash, and I sat at the place of honor on my father's couch so little Gizela could sit with her uncle, and on the other side of her was my father. The tablecloth was laid out with all the food, candles burning, the house was open to neighbors and friends from the street. Vwadzu came over, and Taduscz, Joopi, Theo and his sister Kathe, and Scheng and Leo. Everyone on the street had to see Josko and his American friend. We ate and drank all night. All the coal miners wanted to arm wrestle Hannah because he was so big—and the coal miners weren't office boys who went to the gym twice a week, either. These were men who did a day's work and when they saw the size of Hannah, with the silver stripes on his arm, they all wanted a piece of his action. He loved it, he took them all on, but he never slapped them down, he just held his arm straight up while the other guy pushed and pushed and pushed. You knew if he moved it was all over, but the arm with the stripes never moved. They couldn't budge him.

I'd never had a party at my house before. Now I was the toast of Frankstraat and I was in my glory. My father and I sat on his couch together with little Gizela between us. It was his throne, so to speak, the place he always sat in the house, and all the food was steaming on the table with people eating like crazy, feasting on the food Hannah and I had brought in the trailer. It was like everything else since Big John pulled me up on the half-track and said, "You want to come with us?" It was just a totally different atmosphere from anything I had experienced before at home. It was loose, and some of the frustrations that had always been there were not on my mind. I didn't think about anything that happened before between my father and me. The violence and humiliation was gone. I had a .45 on my hip, my friend was a giant American, there was a Jeep parked out on the street and I could get up any time and walk out the door. My role in that situation had totally changed. It was a transition for my father too. Whatever he felt toward me before, this was a father I never had. I was a kind of person he had never seen before either—I was happy, full of laughs. I was more like the Americans and it must have been strange for him, for everyone who knew me before, to see me sitting there in the kitchen wearing the uniform of a soldier in the American army.

The party went on till morning and my father and I talked about Germany and Yugoslavia and Holland, about the future and what would happen politically. The conversation wasn't personal, but I told him what I wanted—what I hoped—to do. We were getting into the Bols gin pretty heavily, and the effect that drink had on me was to make me feel rather soothing toward him, and when I told him I hoped to go to America, he said, "Josko, please—let us come over there. That's the one thing I want to see."

My aunt heard us talking and she said the same thing, pleading with me to take them to America. "Josko, please, I'll wash your clothes, I'll cook your food, just let us come, please, let us see America."

My aunt had always told me she had a chance to go to America, to Pittsburgh. She called it "Spittsburgh" and if she had known what Spittsbugh looked like—an oversized Heerlen with stinking smokestacks—she might not have wanted to go, although she couldn't have been much worse off than she was in Holland, and probably much better. She had told me that when she lived in Vrbosko someone was going to sponsor her so she could emigrate, then she met my father and married him. She told the story again that night, reminding me she had once had her chance and begging me to take her to America. My aunt, my father, Dragitsa, they were all begging me to take them with me, but instead of feeling that now I had something they wanted and all those years they had deprived me, I had no feeling of gloating. What I felt instead was that now I had a father, a real father, and I had a family.

Everyone knew Josko was going to America and after the visit to #22, driving back to Germany with Sergeant Hannah, I thought the best thing that ever happened to me was the war. Had it not happened my father would never have let me go back to Yugoslavia, he would not have gotten me a passport, he would not have paid for my ticket. I would have been stuck in Heerlen, and my destiny would have been to stay there and work in the coal mines.

Chapter 9 Wetzlar, Germany; March 1946

After the war in Europe ended in May, the army set up the point system. If you had enough points—I think it was 120—you got discharged and sent back to the States. If you had less than 100 points you were shipped to the Pacific because the war against the Japanese didn't end until August when they dropped the A-bombs. If you had between 100 and 120 points, you stayed in Germany until the time came that you had accumulated enough points and then you were discharged. They added points for the number of weeks or months you'd been in combat and most of the guys I was with had been in so many campaigns they had more than enough points to go home as soon as they set up the system.

One day during the summer after V-E Day, Captain Hardesty drove me to the big army headquarters outside Frankfurt, in Bad Homburg, about forty miles from Wetzlar, where we filled out the papers to get me officially inducted into the army. There was a procedure for this because many GI's had picked up a kid, a mascot, and then got him into the army so he was sponsored to go to the States. I met some of them in other units, usually Polish or Russian or Yugoslav, because they were the ones who had most likely been separated from their homes. It was easy for a Dutch or French or Belgian kid to get home, but not so for the Slavs. In the meantime, while the papers were in motion, guys were rotated out and new guys brought in to replace them. We got a lot of guys from the 100th Division who didn't have enough points and the army was dissolving their outfit. Then we got a new group of officers, including a new captain. Captain Hardesty was going home. Before he left, he said to me, "I'm sorry we haven't heard from headquarters yet, but you should hear before too long and then you'll be officially in the American army." The papers were filed, it was just a question of waiting for word they had gone through. After the captain left, they also shipped out Sergeant Hannah. I was very sorry to see him go because we had become such companions. It was always "Sarge and Little Joe are back with the booze," or "Wait till Hannah and Little Joe come back from Holland," and now he was gone. Roper had gone, Riley had gone, all my good friends. I was still living in the small house on the far side of the square in the convent, so after Hannah left Gingold came to me and said, "Joe, I'm going to take Hannah's room, do you mind?" Gingold was a sergeant first class so he had some priority, and I told him it was okay with me, I didn't need the whole place to myself. I got along with him fine, although we weren't buddies like I was with Hannah. Gingold moved into the living room and one day he was lying on his bunk and he said, "Joe, what would you say if I took you back to Holland?"

"What would I want to go back there for?" I said. "I want to stay here. My papers are going to come through pretty soon and then I can go to America."

Everyone had talked about it as if it was a 100% sure thing—Hannah, Captain Hardesty, Big John, everyone, "Little Joe's going to get in the army. No problem."

"I don't think you will, Joe," Gingold said.

"Captain Hardesty put in the papers," I said. "They'll come through."

"I'll bet you anything they don't," Gingold said.

Gingold was just trying to help me. He worked in the orderly room and had read the papers I submitted. He knew my history, that I'd been in a labor camp, and now he was trying to help me get back to Holland because he'd gotten the gist of what was going on in the orderly room with the new captain. Gingold didn't want to see me get stuck in Germany because it was not a good place to be at the time. He was trying to do me a favor without actually saying my chances were zero. But he also didn't know that if I went back to Holland I'd end up in the coal mines.

It was winter but the daily routine living at the convent was still very pleasant and things were easy for me. I wasn't that worried while I was waiting for my papers to come through, although I was getting anxious for it to happen. Then I got a letter from my sister. I hadn't seen her for awhile because now that Sergeant Hannah had gone back to the States we weren't making runs back to Holland for booze, but she had my address, she knew where I was, so I wasn't surprised to hear from her. The surprise was what she said in the letter. She had never written or said anything to me like this before. She began by saying how glad she was I had come to see the family at the house in Frankstraat and also said how fond she was of me. It sounded very strange when I read this because she had never expressed any feelings toward me before. Then she said she had seen Kathe and wrote, "She is wondering what your intentions toward her are." Kathe was a very pretty girl who lived in Heerlerheide and I was in love with her—I hadn't forgotten Basha, but I was in love with Kathe too—and I had stopped by her house to see her on one of the trips back to Heerlen with Sergeant Hannah. Afterwards we had been writing to each other. Kathe was special to me not only because she was so attractive but also because of her parents. The Milicic's were Yugoslavian and when I had my first communion they were the only ones who came to the house. They brought me a present too, a comb in a little sheath of cardboard. They came uninvited, just to honor me, and it was a great surprise and also very embarrassing to my aunt and my father because no preparations had been made. There was no party, I was outside in the garden working. My first communion had been put off for two years because my father wouldn't buy me a suit to wear. Finally the priest said, okay, you can walk in the back of the procession. Naturally it was a humiliating experience because it was just at the age when you begin to be aware of girls and you want to look good and instead of a suit I was wearing the same clothes I wore every day, shorts with a rip up the side, and I was bigger than everyone else. When the Milicic's came to the house they also brought their daughter Kathe who was about eight, and by now the little girl had grown up. We were still both very young and I don't think I would ever have married Kathe, but nevertheless I did love her. She was someone very special to me. So when I read Dragitsa's letter asking me what I was going to do—what are my intentions toward Kathe?—I suddenly felt all my anger and frustration about the family. Maybe she felt that because I had stopped at #22 Frankstraat with Sergeant Hannah that all was forgotten, but I couldn't forget. Especially when Dragitsa was writing to me as if she had a right to question me. As if now she was my real sister and I had always been part of the family. Was I supposed to assume that nothing had ever happened? That my life at home had been a nice thing, like anyone else's childhood? I wrote her a long letter, a very long letter, reminding her of what my life had been like at home. I summoned up my father's treatment, how everyone knew all the beatings I got and the night the Dutch authorities came to the door and I slammed the door in their face. I even wrote down the time he beat me so bad I shit my pants. I put everything in the letter that had been on my mind for years, including that my father had not given me his own name. The family was Merkas, I was Filipovic. These were things I should not have written to anyone, but I couldn't help it. What Dragitsa said had broken this dam and the feelings of all those years came out.

I didn't know her husband's address in Den Haag, so I sent the letter to Dragitsa at #22 Frankstraat where she was still living during the week. It wasn't long before I got a reply from her—a blistering letter saying her father was very upset. He and my aunt had opened the letter and she had read it to him, before Dragitsa came home. Soon after I got a letter from Kathe. She said, "Josko, your father came to the house today and he was in tears. He showed me a letter you wrote to Dragitsa and told me I should break off with you because you are no good. He said you can't be trusted."

It really hurt me to read that. If he had ever felt for a moment I was his son—now he denied it by saying I couldn't be trusted. I felt as if I had destroyed his illusion of me. When I went to the house with Sergeant Hannah and he hugged and kissed me, perhaps he thought now all is forgiven. As ugly and miserable and vicious as my father had been, he might have believed in his mind he did the right thing the way he treated me. Then he read my letter. He walked that long, long distance to Kathe's house with the letter in his pocket, and cried when he showed it to her.

I never should have written the letter. It didn't accomplish anything and I disappointed everyone. But when Dragitsa wrote to me as if nothing had ever happened, it all came back—all those years no one had wanted to be my sister or father or mother—and now it was my turn to say, "You're *not*." I lived with the feeling that I had no father or mother, that people like Hannah and Roper and Riley were much better to me than my family ever was. Writing the letter was not a spur of the moment. Somehow things reach a climax. So maybe what happened was right, and my father did exactly what he should have done by reading a letter written to Dragitsa. It's the kind of thing I never would have said to him and because it wasn't meant for him I was very free in my bitterness. It wasn't only the bitterness I felt, it was the sadness. It was sad because my father and my aunt killed my mother. They were the ones on trial for that, not me. They had always put me on trial for being the bastard but then when I came home as Little Joe, wearing the uniform of an American soldier, suddenly I was the one who was fortunate and their lives were so tragic. "My life has been so horrible, Josko, please take me to America with you!" my aunt kept saying the night I came home with Sergeant Hannah. Her life was horrible! Why? Because of what her sister had done, she'd had to live with this bastard! The one thing I had said in my letter to Dragitsa was how my father cursed me every day with that word, and he cursed my mother for seducing him. The older I got the more I despised him for saying that. I never believed him. Never for a moment did I believe him. But even if she had seduced him, she paid the price. I paid it too. My mother and I were both wrong. She was wrong for what she did, and I was wrong for being born.

The next year my father died. When Dragitsa wrote to tell me, I thought of the letter. I felt as if I had killed him. I had wanted to kill him, a thousand times, and I had never felt it was wrong. But now I felt as if I really had killed him. I had written all the things he had done to me and it literally broke his heart.

The new captain who replaced Captain Hardesty was fresh from the States. He hadn't been in the war at all, sat it out stateside, and now he was in Wetzlar enjoying Captain Hardesty's set up. He had his bedroom and the two mistresses who went with it, the singer and the other fraulein. The new captain was of German descent and was a pretty goddam arrogant clown. There hadn't been any problem between us, he just had no use for me.

A few weeks after my conversation with Gingold when he volunteered to take me back to Holland, the orderly came in and said, "Little Joe, the captain wants to see you. Pack everything you've got in your duffel bag and bring it with you to the orderly room."

"What's going on?" I asked the orderly.

"I think he's going to kick you out."

"Shit, where am I going to go?"

"He called up the DP camp to see if you could go there."

There was a DP camp in Wetzlar; there were still DP camps all over Germany filled with people who couldn't go home or had no place to go.

"Man, I don't want to go to another camp," I said.

"What the hell are you going to do?" the orderly said.

I packed my things in my duffel bag and went to the orderly room. When I came in, the captain said, "Empty your duffel on the floor and show me what you've got."

I had uniforms, fatigues, OD's, my nickel plated .45, coffee, toothpaste, underwear, socks. I had nothing of real value, just personal things, because whenever I got anything like a camera I always gave it to one of the GI's. Because of what the GI's had done for me, I always felt like I wanted to give them something back, like finding frauleins and booze. My attitude was that I couldn't do enough for them.

"All right, Filipowic," the captain said. "Leave everything here except two pairs of socks, two pairs of underwear, and the clothes you have on. Take off the patch—"

I had on a patch for the 29th Division and I ripped it off.

"—now get out."

He meant it. Get out. Get the fuck out.

I felt pretty bad. One of the guys drove me to the DP camp in Wetzlar where an Englishman in a British army uniform shook my hand and greeted me.

"Welcome, we have a place for you to sleep and we'll take care of you."

He had the insignia of UNRRA, the United Nations Relief and Rehabilitation Administration. He was friendly and helped me fill out the form, my name, where I was born, then a Polish guy walked me across the parade grounds to another building. On the second floor was a large room which had been a German army squad room with about ten beds and two cribs for children. There were three married couples, all Polish, one with two children, and the rest were single guys like me, one Polish, another Yugoslav, and an Italian. There were German army lockers against the wall and the beds were all around the edge of the walls in front of them. There was a big table in the center with chairs, hot plates all over the place, canned American food stacked up on shelves, clothes hung up drying, kids screaming. It was a mess, a depressing place, and the people had been beaten down by everything they had been through. There was a bed with a mattress on it, folded over, and the Polish guy who'd brought me over said, "Why don't you leave your things there and we'll get you some sheets and blankets." I got a pillow and bedding from a supply room, came back, made my bed and sat down, feeling dejected. The other people asked where I'd come from and they cursed when I said I'd been with the American army and a Kraut-loving officer had kicked me out.

Things were changing in Germany. It was late winter now and the armistice had been the previous May. The Americans had formed the Constabulary Police which was made up of MP's, Military Police, and they were starting to push the DP's around. The non-fraternization policy was over, the Germans started to mingle with the American GI's, who by now were almost all new replacements who hadn't fought in the war. When the Germans complained to the Constabulary that "the filthy Polacks" or "the filthy Russians" had robbed or raped or killed somebody in the village, the police were more sympathetic to the Germans than to the DP's. It was a continuation of the same friction between the Germans and the DP's that I'd seen when I was with Captain Hardesty and Sergeant Hannah—only now I was on the other side. The *Fräuleins* especially would say, "Look at those filthy *Ausländer*." Whoever they were, Polacks or Russians or Yugoslavs, they were always "filthy," and whenever I heard it I thought of the way we were treated walking through Kohlscheid on the way to the coal mine, regarded by the Germans who lived there as something disgusting. The Germans by now were established and very obliging in the way they did jobs for the Americans, repairing their trucks, doing their laundry, working as waiters in the mess halls. Things had swung the other way. Because the replacement troops hadn't been in the war, they had no idea what the DP's had been through, and had no love for them. And vice versa. To many DP's, the Americans were the Constabulary, the police.

Before I left the office where the GI dropped me off, the Englishman said, "If you want a job, come to the office tomorrow. If you work you'll get extra rations." The rations were adequate but if you worked you got more meat or extras like dried raisins. I would have worked just to be doing something, extra rations or not, so I went in to the office the next morning and the Englishman asked if I could drive a truck.

"Sure," I said.

"Can you drive a two-and-a-half ton truck?"

"Yes, I can drive it."

"Where'd you learn—did the Yanks teach you?"

I said yes but he wanted to see for himself. We went to the motor pool, got a truck and I drove around.

"Okay, let's get you a license."

My job was hauling wood from the forest. It was March, there was snow and ice left in the forest, and we still needed wood for heat in the camp. I had been dejected by getting tossed out of the army but I started making friends right away, and I was still a pretty happy-go-lucky guy, making the best of it. Right away, everyone made a game of the job by racing the trucks back and forth, and I got a reputation as the best driver. The craziest, too.

I had plenty to eat and the barracks were well-heated, the discomfort was everyone living in the same room together with kids crying all night, the married people screaming at each other—it was a hell of a situation if a guy wanted to make out with his wife with all the people in the room. It was embarrassing for them, but they made babies and babies were born in the room.

It was probably more frustrating for the others than for me, although everyone felt the tension. I came in one night and the single Yugoslav guy, an older man about fifty, was drunk, laid out on his bed, and the Italian in his Italian army uniform was beating him. The Italians had invaded Yugoslavia along with the Germans during the war and at one battle the Yugoslavs took no prisoners, so there was no love lost. I walked in on this scene and the Yugoslav was unconscious, his head bleeding. Some of the Poles were standing around watching and I asked what happened and all they knew was the guy came in drunk and the Italian started beating him. I went over, the Italian saw me and started to say something and I hit him under the left arm, dislocating his shoulder. He started screaming like a pig in pain, so I threw him out the window. Then I threw his bed out the window after him. Also his wall locker. Twenty minutes later, the camp police—they were Polish DP's, different from the Constabulary—came in with the Italian. I had dumped him on the pavement but he was all right. He pointed his finger at me, screaming and hollering, and they didn't need to know Italian to understand him. The Poles asked me if I had thrown the guy out the window. The Yugoslav was still out from the booze and by now there were about fifty people around us, coming in from the other rooms. Before I could answer the camp police, the other people all started shouting that the Italian had been beating the Yugoslav and Josko hadn't even been in the room. They said they had scared the Italian so bad he jumped out the window and they threw his bed and locker out after him. They're all taking the blame. The camp police didn't question me at all. Then the English officer came in to see about the commotion and asked, "Does anyone here speak English?"

"Yes, I do," I said.

"What happened?" he said.

"I threw this fucking guy out the window," I said. "And I also threw out his bed and his locker."

In his limey accent the English officer said, "What'd you do that for?"

"He was bragging about how many Yugoslavs he'd killed in the war," I said, which turned out to be a lie. Later I learned it was the other way around, the Yugoslav had bragged about how many Italians he'd killed.

The Englishman was a wiry little sonofabitch, 130 pounds with his boots on, and he smiled and said, "I don't recall the Italians killing anyone."

"Well, he was trying to kill this one," I said.

The officer looked at the guy on his bed, still out—dead drunk—and said in his limey accent, "He doesn't look the worse for it."

Then he looked at the Italian—bleeding from his head, his arm was in agony—and he said, "Matter of fact, he doesn't either."

One of the Polacks translated that and broke everyone up laughing.

Then someone suggested that the officer might be kind enough to get the Italian another place to live, and the officer ushered him out.

There were a lot of Czechs and Hungarians in the camp as well as many other Yugoslavs and whenever I met a Yugoslav I asked if they had come from Vrbosko. I never met anyone who did. Later I heard that my grandmother hadn't survived the war and many of my nephews had died. All my relatives either died or I lost contact with them. It was impossible to correspond with anyone in Yugoslavia after the war and I never heard from my friend Petar. Years later when I went back to Heerlen the Milicic's told me about Petar's family's fate. His sister was killed, his younger brother and his father were also killed. Only he and his mother survived. Petar had joined Tito's partisan army and has risen to the rank of colonel, one of the youngest in the army. During the war I'd imagined him fighting with Tito against the Germans, and that's what he'd been doing.

The gates hadn't been opened yet for emigration and many of the people in camps were just waiting to find out what their fate would be. The Poles didn't want to go home under a communist regime so many of them were still in camps. The Russian DP's had no choice because the government demanded their return as part of the deal at Potsdam, so many of them were forcibly sent back—and then straight to Siberia. Yugoslavs weren't forced to go back and I heard from people who'd escaped that things were very rough there, they were fighting and killing, and I wanted no part of that. But if you didn't go home, where would you go? There was a great deal of uncertainty for the thousands of people who'd been displaced by the war. The food ration we got in the camps was enough to survive but that was it, and meanwhile, during the waiting, everyone struggled to have some kind of life.

Driving back to the camp with a load of wood one day, I was at the bottom of the hill, about to shift into low gear for the drive up the steep hill that led to the camp, when I saw two DP's thumbing. I stopped and told them to jump in. They were Polish guys, about forty, very friendly, perhaps because they'd had a few drinks, and we were speaking Polish when they asked my name. "Josko," I told them.

"Oh—Josko! We're celebrating your name day!"

It was March 19th, St. Josef's Day. I hadn't noticed.

"You've got to come to our party tonight—we've got food, vodka—a big party!"

They told me what building they were in and after I made my last trip for the day I went back to my room, washed up, and walked over. As soon as I got into the building I could hear the noise. They were playing music and singing. I knocked on the door and one of the guys I gave a ride to opened it. He gave me a hug, brought me in—it was a big room like I was in—and introduced me. "This is Josko, he gave us a lift up the hill...." Right away they gave me a porcelain cup and poured vodka in it. It was the foulest smelling stuff—just the smell alone without even tasting it made you dizzy—but I had to drink with them and after only a sip I could feel it. There were several other Josefs celebrating their name day and they gave me a place at the huge table. They had a white tablecloth, plates and nice settings, probably all from some German's house, and they also had a whole pig. Wherever they got it, they sure hadn't raised it themselves! It was just like when I was with Captain Hardesty and Sergeant Hannah, the DP's would go out, grab a pig and run for the woods and the Constabulary Police couldn't control it. But nobody could have run far with this one because it wasn't a piglet, it was a big sow—and still in one piece! Not only that, they had managed to barbecue it. It was beautifully roasted and browned, a real thing of beauty. The vodka wasn't stolen, they had made that themselves, probably the same morning. It was so foul if you smelled it first and then tried to drink it, you couldn't. You had to block your nostrils and throw it down, then shove the pork down your throat. That was the way everyone ate. You got a knife, sat down at the table, cut a piece off the pig, had a sip of vodka, and chased down the vodka with pork. I think the pork is what kept you from going insane, because I was never so out of my mind as I was after that feast.

When the Polish guys brought me home they had to carry me, one under each shoulder, and in the building I lived in there was a hallway from one end to the other, lined with windows, and as they hauled me along I busted every window and frame, BAM! BAM! BAM! The two guys were as drunk as I was and tried to stop me from smashing the windows, but it wasn't easy because they could hardly walk themselves. They got me into the room and put me in bed. I was lying there, I hadn't passed out yet, when a knock came on the door and the camp police came in. It was six Polish guys, DP's, in their dark blue dyed American uniforms and helmet liners carrying carbines and billy clubs. "Who busted the windows?" they asked. I was the only one in the room who'd gone to the party but no one said anything. Someone else had obviously heard the racket and reported it.

"Did you do it?" the Polish guy asked me.

"Naw, I didn't bust anything," I said. I must have knocked out twenty panes but the miracle was that I didn't have a cut or even a scratch on my hands.

They asked the others in the room, mostly Poles themselves.

"No, he didn't do it," they said.

The camp police didn't believe them.

"What do you think, should we lock him up?" the policeman said.

"You better not lock him up, he's a good friend of the camp commander because he speaks English," one of the people in the room said.

It wasn't true. I barely knew the guy.

The DP's talked to each other for a minute, then one of them said, "You better be good."

The next day I had a wicked hangover and when I got to the motor pool to pick up my truck, the Englishman there—a different limey—said, "I read a report this morning. I hear they're replacing some windows over there in your building."

"Oh, yeah?"

"You better be careful, you'll kill yourself." It wasn't a threat, just friendly advice. "How'd you do it?"

"I hit it with my fist."

"Let me see."

I showed him my fist which by now was a little swollen.

"I didn't know you drank," the limey said.

"I don't drink, but I went to this crazy Polack party where they had vodka."

"Vodka?" He was interested.

"I can get you some if you want," I said. "But you don't want to drink this stuff. You don't *ever* want to drink it."

"It's that bad?"

"Worse." I said. "They made it themselves."

The party had gone on all night and there were complaints about the rowdiness but only the camp police got involved so there was no trouble. Had it been the Constabulary Police it would have been different. The friction with them just kept increasing. The fresh, non-combat officers with a big "C" on their helmet liner—they just wore the liner, not the helmet—were doing what the Germans had done during the war, subjecting the DP's to brutal treatment. The "C's" would come in and claim on the say-so of a German that a DP had cut someone's throat. Where Captain Hardesty would say that was the German's tough luck, the C's would pull people out and beat them. I came into the room one day and six C's were beating a Pole. I went after them and when a punk lieutenant shoved a .45 in my nose, I took a very belligerent attitude toward him. "Don't try anything with me," I warned him. "I was fighting with the Ninth Army while you sat out the war in the States." He backed off, the GI lingo got his attention, but he was a typical wiseguy replacement. The new American troops had no idea what the DP's had been through. They would see the shit all over the bathroom and think the DP's were disgusting, they were vegetables. My attitude changed too. When I was with the Americans I was more compassionate. Now I was in a camp again, being pushed around, and the people I was with had already been through this, and much worse, before.

Driving the truck I came down the hill faster than anyone, with the guys in back hanging on for dear life, and because I drove so fast we got off earlier. But the Germans in the villages I drove through complained. They said I scared the women and children and horses. Goddamn right I did! I was going like hell, trying to kill a German a day. I passed on hills, on the right, on the left, and to avoid me the other trucks would drive right off the road into fields. I didn't care, I wasn't going to stop. In the villages I'd chase people right up on the sidewalk, and the guys in the back of the truck loved it. They were mostly in their twenties, mostly Polish, and these men weren't gentle. There was a rumor that some Poles took five minutes to slit a German's throat with a pocket knife. They didn't even fear God. They had been in labor camps but unlike me they had never had the experience of being with the Americans. I'd been on other side where I was treated well and the Americans were great friends. But to these guys the Americans had become the enemy, like the Germans. The DP's had gone from being in camp to being in camp. They might have thought this new camp was a better experience because they could eat, but that was all they could compare it to.

We were in the forest loading the truck with firewood when the American MP's came up to us and said a German had been found dead and the people said some DP's did it. I said, in English, "It couldn't have been these guys because they've been right there with me."

I hadn't seen the Poles kill a German although they could have. I'm sure they would have been glad to.

"So who the hell're you?" the MP asked.

To him, I was just another DP, hauling wood, wearing the clothes. But the one thing I had, besides my ability to speak English and to talk the GI lingo, was a copy of the letter Captain Hardesty had written to accompany my application to officially join the army. It described how I joined the American army during the war and had proved myself to be a good soldier and an outstanding person. He signed it, Captain Dennis Hardesty, 29th Division, Ninth Army, etcetera. I always carried the copy of the captain's letter with me and I took it out and showed it to this eager young guy from the Constabulary. He sort of frowned at it, like he had never seen anything like this before, but whatever it meant to him, he accepted it, and I was able to talk him out of taking us away. Nevertheless, it was the kind of incident that put everyone on edge and where before the American uniform was reassuring, now you had to be careful.

In July they announced they were going to clear us out of this camp and take us to Kassel. It was about a two hundred mile round trip and I was one of the ones designated to drive people and their belongings in the trucks. The new camp was in the woods outside Kassel, a place where they had kept Nazis they rounded up after the war, and when I drove up what I saw was a labor camp if I'd ever seen one—a big gate, barbed wire, barracks, a couple of houses—exactly like the prison camps the Germans had kept us in. It was very depressing. They moved everyone into the empty barracks, families, children, fifty or sixty in a room, and no one had any privacy. The best people could do was hang up sheets. At least the floors and walls were wood and not concrete like at Kohlscheid, but the Americans had let the Nazis go and put in the DP's, and the DP's got the message—now the Germans were the favored people.

I drove back and forth to Kassel for more than a week hauling more and more people, and it was awful taking them to the camp. It was way out of town in the woods, it was rainy with mud everywhere, and the first impression people had was the barbed wire. It scared the hell out of them. Eventually they took it down but it was an ugly shock the first time they saw it.

I had met the Canadian naval officer working for UNRRA who was the top banana in the camp because the Englishman in charge of the motor pool in Kassel told him I was a good worker. I drove like crazy but I always made more trips than anyone else, and I could also translate so the Canadian always called me when they needed an interpreter. Besides German and English I could speak Polish and Russian with the DP's and I had also picked up Yiddish just by hanging around because there were Jews in the camp. Outside the camp was a house with private rooms where the Canadian and the other UNRRA personnel lived and he often invited me to come over. He was single, or had no family with him, and one day when we were talking he said, "Joe, I need someone to help inspect the facilities in camp. I do inspections myself, but I need someone who can talk to the people. I can pay you well, you'll work in an office, and you won't have to drive a truck anymore."

It sounded fine.

So I became the civilian in charge of all the utilities. My job was to walk around every day inspecting things, listening to people's complaints and also tell them what the commander wanted done. Besides the inspections they were constructing new buildings and I just walked around with my hands in my pockets while everyone else was working their asses off. I was living in one of the barracks with about fifty other people so the Canadian put a bed in his quarters outside the camp and I stayed there most of the time. After about three weeks of doing the job inspecting facilities, the Canadian said, "I've got another job for you, Joe. A better one. I'm going to make you the civilian camp commander."

"There are two thousand people here," I said. "You want to make me the civilian commander?"

There were a lot of obviously well-educated people running around and acting important—Poles and Russians—and the Canadian said, "Those guys drive me crazy. I'm going to put you in charge and if anybody wants anything from me, they're going to have to go through you."

The problem for the Canadian was he had such a bad job. Trying to keep two thousand people happy under the circumstances was impossible. He was sincere, and he worked hard, but it was too much. He couldn't get anything done because people were constantly rapping on his door. He thought it would help if I was the one answering all the complaints.

I said okay.

"You see that house?" He pointed out the window. "You can live there."

"That's too much for me," I said. It had four bedrooms, a bathroom upstairs and down, a kitchen.

"Okay you can share it with the assistant camp commander. He wants to be commander pretty bad but I can't stand the guy, so I'll make him the assistant and give him part of the house. You want up or down?"

"I'll take the downstairs."

He called the other guy in and said, "I want you to meet Mister Filipowic. He's the new camp commander."

The guy looked like he was going to faint. He was Polish, about fifty, nicely dressed, looked like a professor, and he didn't like the idea of working for an eighteen year old kid at all! He gave me a look that said what do *you* know? He was right too, because I didn't know anything. All I knew was I could talk to people, I could understand them, and could also make myself understood.

I moved into the house next door, it was all furnished, and the Polish guy moved in upstairs. Somehow in all that mess he had found his family, and his wife and two grown daughters moved in with him. But now he was pissed off because they were cramped upstairs and downstairs I had two bedrooms, a kitchen, and living room, besides a bath. I had met a girl named Elena at the camp in Wetzlar, and she was always complaining about the living conditions because it was so miserable in the barracks. There was no privacy, people were always fighting and screaming, so I said why don't you move in with me. She came over to the house, took one look around and said, "God! Of all the people who could have gotten this job, how did you do it?"

I said, "I got the job, now shut up and start cooking."

She said, "I'd be glad to," ran back to the camp, picked up her bundles, and moved in.

Elena was a young, pretty girl who had been picked up in Poland during the war, put on a truck, and shipped to Germany to work in a factory. Now she was like so many other DP's, sort of floating with everyone else in the chaos. She didn't know what had happened to her parents, but she didn't want to go back to Poland. She was happy because of my getting the job and the house, so for the moment things had worked out for her. But it didn't last long. The job started driving me crazy too. I had the same problem as the Canadian—it was constant, constant, constant complaining, and I was like a ten year old put in the position of being an executive. It was a catastrophe.

Everyone had suffered so much and here they were in another camp. Of course they complained, living like this made them crazy. In Wetzlar, the Constabularies had just come in one day and said, "Everyone out! Out! Out!" It was like the Germans had treated them when they were just dragged through the streets, and raped and abused, and now they had been put in a camp in the woods surrounded by barbed wire. And who were these people? Polacks, Russians, Yugoslavs. No Dutch, no French, no Belgians, just the Slavs. All the Nazis were let go and the Slavs were put in prison, the same prison they had just let the Nazis out of! And everyone knew it! So for these people the liberation was something that still hadn't happened. The war was over but they were back in a camp and the Nazis were free. The Canadian had compassion for them, and he was even perhaps trying to help me as an individual by giving me an important job, but he wanted me to be something I couldn't possibly be. When he said, "Joe, I want you to organize this, get sewerage there, make sure the kitchen is sanitary and no one is stealing food, and you can do this because people understand you and you understand them," he meant well, but he just didn't know what the hell he was talking about. The one piece of advice he gave me was, "Joe, don't shave with cold water, if you do your face is going to bleed." He was trying to be helpful, but that was the best he could say: "Don't shave with cold water." How the hell was I going to tell these people, "Don't shave with cold water"? There *was* no hot water! I was supposed to give everyone advice, but what the hell did I know?

Everyone loved me but I was uncomfortable. I couldn't do it. I was not authoritarian. I had my own house—in a DP camp!—when other people were living fifty in a room. Elena came over every night to sleep with me, maybe not because she was in love with me but because that was a better place for her to sleep. The job, the situation, was driving me crazy, and I thought what the hell do I need this for? I was eighteen years old. I wanted to dance!

CHAPTER 10
ALSFELD, GERMANY; AUTUMN 1946

On one of the trips back to Wetzlar to get another load of people when I was transporting them to the camp in Kassel, an oil line broke in the truck. I unscrewed it and stopped a jeep driving by and told the lieutenant I'd busted an oil line and wanted to find a motor pool to get another one. The lieutenant looked at me standing on the road beside a two-and-a-half ton army truck, wearing the blue-dyed fatigues of the DP's, and talking like a GI, and he said, "Where'd you learn to speak English like that?"

I told him I'd spent some time with the American army during the war.

"What outfit?"

"The 29th Division."

"Jump in," he said. "I'll give you a hand."

While we were driving he asked me where I was from and I told him I'd been born in Yugoslavia but had lived in Holland.

"You speak Yugoslavian?"

I told him yes, I spoke it. Serbo-Croatian was my first language.

"I've got thirty Yugoslavian mechanics working for me in Alsfeld," the lieutenant said, "and they're the best mechanics I've got, better than the goddam Krauts, but I can't understand them.

If you get tired of working at the DP camp, why don't you come and work for me as my interpreter."

He was a real spit and polish young second lieutenant with his .45, and must have just gotten out of ROTC because he still had the gold bars. He drove me to a big former German aircraft factory outside the town of Alsfeld which the Americans were using as an ordnance motor pool to repair heavy trucks, tanks and artillery. The lieutenant wanted me to meet the Yugoslav mechanics. They were working like hell, hanging all over the trucks and tractors and tanks, dressed in US army fatigues but wearing their Yugoslavian army caps. The lieutenant introduced me to a couple of them and they spoke no English at all, only a little German, so I spoke to them in Serbo-Croatian and they showed an interest because they could see I spoke good English and was fraternizing with the lieutenant. After I met the mechanics he said, "I want you to meet Lothar."

He took me to the mess hall where we met an enormous guy with a pair of hands like sledgehammers. Lothar was another Yugoslav working in the motor pool and the lieutenant told me to ask him, in Yugoslavian, how his countrymen were treated here. The lieutenant walked away while Lothar and I talked, and he said they treated him well, they ate the same food as the GI's, and were respected and liked by the commanding officer. So I got a tour of the motor pool from the lieutenant and also got a new oil line for the truck, and when the job in Kassel got to be too much, I remembered the offer to work for him.

I got a ride from Kassel down to Alsfeld, found the lieutenant who was glad to see me, and started working for him right away. The job was just what he'd said. I translated for the mechanics and the GI's, and wrote out the work orders in Serbo-Croatian. I didn't know the words for all the technical terms but that was something I could pick up, and things went pretty smoothly. We all went through the chow line with the GI's and I ate at the same table with the Yugoslav mechanics. We got along all right, although they didn't take me in with open arms. There was a caution, almost a caginess, in the way they regarded me, partly because I didn't speak perfect Serbo-Croatian but at the same time I spoke very good German and English, so they wondered where I had come from and asked me questions—when did you leave Vrbosko, what were you doing in Holland? They were tough, battle-hardened soldiers who had been captured when the Germans invaded Yugoslavia in 1941. As POW's they had been pushed around by the Germans, although because they were skilled mechanics the Germans had put them to work in factories instead of down in some coal mine. They were proud men and when work finished at night, they cleaned up and put on their Yugoslavian army uniforms with all their insignia to wear home. All of them drove motorcycles which in those days was unusual and they lived in private homes in Alsfeld, not in the DP camp. They were paid, they got three meals a day, the same food as the GI's, and extra rations from the mess hall to take home. They were also given cigarettes which gave them something to barter with, so this was a very good job for them.

I'd been there a couple weeks doing interpreting for the mechanics when the lieutenant asked me if I'd mind taking one run a day picking up Germans to bring them to work and then dropping them off at night. There were at least two hundred people working in the motor pool and they were short on civilian drivers. They preferred to use the GI's for other jobs instead of driving and the Germans lived all over the place in small villages. For a young kid like me driving a two-and-a-half ton truck was fun. Every morning I picked them up and every day I gave them the ride of their lives, racing from the top of the hill, double-clutching down the mountain. They rode in back dying with fear but were afraid to complain about my driving because they all needed the job so bad.

One of the Germans was a former POW named Fritz working in the motor pool who kept bugging me to teach him to drive. "You were in the German army, didn't they teach you to drive a truck?" I said to him.

"I was in the army but I never even *sat* in a truck," he said. "I was always marching."

"I can't teach you to drive. You have to get permission."

I didn't like the guy because I could see that underneath he was really a nasty arrogant German. He was on the fringe, not nasty enough to antagonize me to kick him in the ass, but nasty enough so I could see what he was.

Fritz had been sitting in the truck every chance he got shifting gears, working the clutch and the gearshift, only of course the truck wasn't moving, and he kept bugging the motor pool sergeant to give him a chance. Finally the sergeant said to me, "Joe, this guy claims he can drive a truck. Will you take him out and see if he can?"

"The sonofabitch can't drive—he told me himself he can't," I said to the sergeant.

But the sergeant was willing to give him a chance because he needed drivers.

Fritz got behind the wheel of the truck I drove every day and we started down the road by the railroad tracks, past the huge hangars. He drove down the street holding on to the wheel, not doing too bad but he was still in first gear, he hadn't shifted yet. At the end of the road he had to make a right or a left. He made a left. But when he made the turn at the corner he turned too late and didn't quite make it, so he steps on the brake and stops about two feet from the building across the street. It's an office building for the ordnance depot. I can see people inside through the window, typing, doing paperwork. Now Fritz has to back up. He gets a little nervous. He steps on the clutch and puts it in gear. I noticed that he had taken it out of first and put it back into first instead of reverse. He's got his hand on the wheel and twists his head around so he's looking through the back window. He lets out the clutch and steps on the gas. I looked through the windshield and wondered which window the truck was going through. There was a jerk as we moved and then WHOOOOM! We went through the wall—right through it into the office. The winch on the front of the truck was like a ram. It made a big hole, then the bumper went through and the rest was easy. It pushed desks and partitions around, bricks were falling, and the truck was open so I rode into the office with my arms thrown over my head. Women were screaming, people dashing in every direction. When I got out I was standing in the middle of the office. Fritz still had his *Wehrmacht* cap on—dyed blue—covered with concrete dust, and in the middle of all the commotion, the sergeant from the motor pool came running up and said, "What the hell happened here?"

"This stupid bastard was trying to back up," I said, "and instead of putting it in reverse, he put it in first."

"He put it in first? I told you to take him out for a test! You must've seen that, Filipowic!"

"I thought he'd just bump the wall," I said. "I didn't think he'd drive right through it."

"The hell you did—even if he was only going to bump it, you should have stopped him."

"I told you he couldn't drive," I said. "You're lucky. He could have hit something else and wrecked the truck. This way it only got scratched."

"What about the goddam *building?* What're these Krauts going to do? They're supposed to be working for us!"

"Well, you've got a point there," I said.

"I'm going to get your ass in front of the captain!"

While the sergeant was chewing me out, the local German police drove up. Right away they set up to interrogate Fritz. They're going to punish the poor bastard, how I don't know.

I backed the truck out of the office and drove it up to the motor pool where everyone was laughing like hell and one of the Germans, another POW from the east, a crazy guy, was already composing a song about Fritz at the wheel. "Fritz takes a drive in the truck, Fritz puts the gear in reverse, Fritz and the truck go forward, right through the wall..."

The gang in the motor pool I saw every day was Fritz, Gipsarsch and two mechanics, former German sailors, one from Bremen, the other from Hamburg. The sailors had both been wounded, sent down for convalescence, and the war ended while they were recovering. Neither wanted to go home because there was no food in Bremen or Hamburg. The other German, Gipsarsch, was local, a farm boy from Alsfeld. He wasn't a mechanic, his job was to wash trucks, grease them and change oil. "Gipsarsch" meant "plaster ass" and the other Germans in the motor pool gave him the name because the poor guy was tall and lanky, a good six feet, but he had a club foot and the way he stood his ass always stuck out like an ostrich. Within a few months he was known in the village as "plaster ass." He was a little slow but very nice, a heart of gold, always trying to be helpful, and unlike Fritz everyone liked him.

The GI's driving the trucks had nothing to do except for a couple hours in the morning and evening, so they were always fooling around and the fad became putting big mud flaps in back of the tires. If anything, all they did was stir up even more dust, but the drivers all wanted them and one of the GI's told Gipsarsch if he could find some rubber flaps for his truck he'd give him two packs of cigarettes. Gipsarsch said, "Don't worry, I'll find them for you. You'll have them this afternoon."

I overheard them talking and thought Gipsarsch must know where there's some rubber lying around. The whistle blew, it was lunch time. We got an hour break. It was a nice, sunny day, and everybody was outside eating. I was sitting on a ramp at the back of the building talking to the two German sailors, and all of us were still laughing about Fritz because it had only happened the day before and the masons were over there putting up a new wall. All of a sudden we heard a loud clap, like thunder. The sound echoed through the whole building and we stood there stunned. Then we heard someone screaming, OOAAAAOOOH! We rushed into the building. Inside were stalls where the trucks and tanks were repaired and way up above them was a huge conveyor belt they had once used to transport sand for casting when it was a factory during the war. Doubled up these belts were over a hundred feet long and under a great deal of tension. Gipsarsch had promised the GI he'd get him rubber flaps, so he had climbed up on top of the conveyor belt with his knife and went VROOOP!—he told me later, "It cut so easy!"—and when he cut the conveyor belt, it flew out from under his legs. He dropped between the rollers but stuck out his arms and grabbed them as he was falling—only they were rolling while he tried to hold on. "My arms want to slide away!" he told me. When we ran in we saw Gipsarsch hanging twenty feet in the air, frantically pushing his hands back over the rollers again and again. Someone had the presence of mind to put up a long ladder and they got Gipsarsch down. He wasn't hurt but the conveyor belt was ruined and because the Germans are great for being official about everything, back came *Herr* Police Chief, and also the captain, and they wanted to know what possessed Gipsarsch to cut the conveyor belt. Gipsarsch was too simple to make up a story, so he told the truth, and the captain stood there shaking his head saying, “I want to go home. I can’t take this. The guy cuts up a $10,000 conveyor belt so some simpleminded ass can put mud flaps on his truck!”

Meanwhile, out comes the pencil stub and the poet begins writing a new song, this time about Gipsarsch and the conveyor belt.

Fritz and Gipsarsch were good for a laugh but the guys I spent time with were the two German sailors. They were five or six years older and made a great effort to befriend me. It was just personal, we got along. One was Max, the other was Johnny, a typical romantic waterfront name, the same in German as in English. In all the German sailor’s songs the guy is always Johnny. They made me feel at ease when we went to the German dance hall in Alsfeld. The Poles and Russians, the other *Ausländer*, didn’t go to the German dance halls but I could pass for a German because of speaking the language like a native. They didn’t know my last name, I was “Joe”, after I got picked up by the Americans I was always Joe. Dancing was very important to us, perhaps because people hadn’t been allowed to dance during the war. Only the elite could dance, not the common people, it had even been forbidden in Germany, and dancing was always a cheap form of entertainment for poor people.

I had plenty to eat and the job was easy, but by the next spring I was getting restless. I’d spent the previous Christmas at the GI service club, which I had access to, but it was also a bore, seeing all the same girls all the time and most of them were camp followers, hanging around the GI’s. Word had gotten out that I wasn’t a German and the girls in the village were afraid to go out with me. A nice German girl, what did she do with a Yugoslav? It just wasn’t done. It was okay for the nice German girls to go out with an American, although for the GI’s they were just shacking up. It was a continuation of the way things had been at Wetzlar, the girls were laying them for food and a good time at the club. Things were still unbelievable for the average GI. They had cigarettes, candy, drinking, and dancing, and the Germans had lost so many young men that it was very tough for the young women to get a man. Nor did they have food. They were still on starvation rations. For a pound of butter or a can of coffee you could keep a girl in an apartment for a month. If a guy didn't have a date, all he had to do was go over to the service club and get one because they were always there. It was unbelievably open, a free-for-all, and the GI's were reaping the benefits of the conquerors.

Johnny and Max often talked about going to one of the big cities like Frankfurt, so I had begun thinking about leaving too. In the spring I went to the lieutenant who had hired me and told him I wanted to leave. He understood and helped me get a job at another American operation in Oberursel. It was a suburb of Frankfurt just a streetcar ride from the center of the city and the situation was similar to Alsfeld, they needed someone to translate in the motor pool. So I became in charge of the parts supply room, just because I could speak English and German. It was an easy job, the pay was good, I got three meals a day, and I could get into town where the action was.

One Saturday I was in the Oberursel *Turnhalle* to go dancing when I saw a guy standing at the bar waiting to get beer who looked familiar. I walked over and said, in German, "Do I know you?"

He looked at me and said, "For Christ's sake—Little Joe!"

I said, "Yeah, that's me. But who are you?" I still couldn't place him.

"I'm Albert Schwänzer."

Albert had been one of the thirty-five German POW's we had locked up in the church at Wetzlar when they were making out with all the frauleins! The POW's had all known me as Little Joe and although I never talked to them much—I never had any desire to—Albert was very friendly and asked me to come and sit at his table. There was a girl and another guy there named Franz who was from the same city as Albert, Stettin. Albert was working as a welder in the same shop where I ran the parts department. He was a couple years older than me and had been in combat in Holland when the British and Polish airborne jumped into Eindhoven, then he wound up near Wetzlar when the Germans surrendered. Although he had been in the German army he had no interest in the Nazi Party or the German *Reich*. He had never joined the Hitler Youth because he had long hair and wouldn't cut it. Albert was a free spirit, a very unusual guy, and we started hanging around together.

Besides the job at the complex, I was working as a waiter at the GI service club in Oberursel. No one took me all that seriously as a waiter—they treated me more like one of the boys—but I got paid, got more food and free drinks and cigarettes. I helped translate for the GI who booked the acts and the bands, and also found him an apartment for his girlfriend, so once again I was operating, mixing, just like at the motor pool where I began to develop a clientele. GI's would come to me around the first of the month when they had to pay the rent for their fraulein's apartment and needed some *Deutschmarks* and I would sell their goods for them. They had cigarettes, coffee, butter, the usual commodities, and occasionally something special like a camera, so my parts department at the complex had a little grocery store and tobacco shop too. I rented a nice room in a house in Oberursel and could bring girlfriends into the GI service club and it was a very pleasant time, a lot of fun, a lot of laughs. Albert and I raised hell together, we went dancing and partied and got in fights just for the hell of it. One night I got stabbed with a broken bottle and bled all over the checked sports coat I'd stolen from the house during the air raid when Leo and I escaped from the labor camp. I'd carried it with me all this time. It was good quality—the German guy must have paid a lot for it—and I cleaned it up and kept wearing it. We were just young guys floating with the madness of the time, enjoying ourselves, and in the circumstances eating and living pretty well.

The one thing on Albert's mind was his family. His father had been drafted by the *Wehrmacht* too so they lost track of each other, and Albert didn't know if his mother had been killed or if she had gotten out of Stettin before the Russians came in. He had been checking with the Red Cross in Frankfurt because they were active in locating people and one day he got word from them that his mother was in Hanover. It was in the British zone, about four hundred kilometers from Frankfurt. Albert wanted to go see her and asked me to come with him. We both took ten days off work and left on a weekend.

The train was so packed that people were jammed in the hallways outside compartments, hanging out of windows, they were even riding on top of the cars, just holding on. It looked like it was going to be a rough train ride, if we could even get on. But they had special compartments for Allied personnel, so we went over to the stationmaster and told him we were with the military government and would he kindly assign seats to us. We showed him the working passes we'd been issued which had our pictures on them. On the front it said "Military Government" in both German and English, but we didn't open it up to show him where it was stamped—and he didn't either, otherwise he would have learned we worked at an ordnance depot—so we got a beautiful compartment and traveled like gentlemen. When the train left the American zone and entered the British one, everyone had to get off the train so they could check the compartments for black market contraband. Baggage was searched, they looked in whatever you were carrying. Albert and I were loaded down with cigarettes and canned food because we figured his mother would need it, so when the train stopped at Eschwege and everyone got off, we just sat in our compartment looking out the window at the American MP's searching everyone. When they came to our compartment and asked for our papers we gave them the same "military government" bullshit and they bought it. The guy just glanced at our working passes and handed them back. So Albert and I had a very nice trip to Hanover.

There was a big scene when we got there. Albert's mother was a little German woman and very excited to see him. He had no brothers or sisters so he was her only child and I could see they had a very good relationship. The first thing she told him was that she'd found his father, he was alive. She hadn't seen him yet but he'd survived the war. Hanover had been badly bombed and was a mess. It looked like Frankfurt except the soldiers were British instead of Americans, and everyone had the same problems, lack of food, no clothes. Albert's mother was just barely surviving and when we arrived with cigarettes, she immediately took a pack and went into town to trade for food. She came back with meat and bread and cooked a big dinner for us. There were other people there from Stettin, friends of the Schwänzers, and they all came over to see Albert. Stettin was a long ways from Hanover, in the east, but when the Russian armies advanced a lot of people just packed up and left, afraid of what the Russians were going to do to them.

Albert and I made a couple trips to Hanover and the next time we went his father was there so it was another big reunion. He'd been in combat and had been a POW but he was okay. The family decided to stay and make their home in Hanover because by now Stettin was part of Poland and they didn't want to go back. His father was a pretty good guy, a lot of fun like Albert, and the family accepted me. I was Albert's friend and they didn't have any of the attitude—what I thought of as the arrogant German attitude.

My legal status at this point was zero. The only thing I had to establish my identity was the working pass from the motor pool and the letter from Captain Hardesty. But the letter hurt me too because it said I'd been picked up by the American army in Holland. When I went down to the office in Frankfurt to see about emigrating—I was willing to go anywhere, Canada, Australia, Venezuela, if I couldn't go to the States—they told me I wasn't a Displaced Person and didn't qualify for DP status because I had joined the American army voluntarily. Even though I was a Yugoslav, my country of origin was Holland. Had I been picked up by the Americans in Germany I'd have been a DP because I'd been taken there by the Germans. By having escaped from the camp in Kohlscheid and gone back to Heerlen, I had repatriated myself and therefore I didn't qualify.

The war had been over for more than two years but there were still DP's and camps everywhere with thousands and thousands of people, all just waiting to see what would happen to them. In one neighborhood in Frankfurt called Salzheim, the Americans cleared out all the Germans and moved Jewish refugees into them until they could be repatriated or resettled with relatives or sent to America or whatever country. The

Germans talked with great disgust about how the Americans had put "all those filthy Jews" in their homes. Since I could pass for a German myself, there were times they assumed I was one of them so I often heard what they really thought. Every time they saw a DP, behind his back they'd say, "That filthy Polack, why doesn't he go back where he came from." For some reason they had an antiseptic feeling about themselves, a superiority, and their attitude hadn't changed just because of losing the war. They were sorry about that, very sorry about the way Dresden got bombed, they thought that was very inhumane, but they hadn't thought so when they were blasting the hell out of Warsaw for twelve, fourteen days with *Stukas* and heavy artillery, and broadcasting *Sondermeldungen* about how Warsaw was being destroyed by the glorious German army. My bitterness was always there because I knew for many of them, if they knew who I was, I was still the *Untermensch*. You could be president of the country or a great scholar and you would still be an *Untermensch.* This applied to all the Slavic peoples and to the Jews, and that was the concept behind killing them. Had the war gone on another couple of years there probably wouldn't have been any left, but they could only kill so many which was part of the hysteria of increasing the production of extermination. It was no secret what was going on. We knew when they rounded up the Jews in Heerlen in 1941 that they were going to the concentration camp. We knew once you went there you would be killed. In prison camp in Kohlscheid, the German guards used it as a threat to make us work. If you were sick they'd say, "Don't be sick Friday," and when they threw you on the truck to go to the boxcar you weren't being taken to the hospital. You went to the *KZ*, and the reason you went to the *KZ* was to die. That's why it was there. Every German knew about the concentration camps. After Albert left Oberursel, I lived in a woman's house whose husband had been in the *Wehrmacht* on the Russian front. She had gotten notice he was lost and assumed dead. Once we were talking and she got out some pictures her husband had sent home from the front. They showed six or eight metal racks with people hanging from them, six people on each rack, dead, with German soldiers standing beside them. They were wearing the black clothes of Orthodox

Jews with a sign that said, "Jew." This was before the gas chambers, when they were hanging them, and he was sending her pictures of what he was doing. She had several of these photographs and they weren't secret. She didn't pull them out of a dark corner or hide them in the attic. They were in a bureau drawer with letters from her husband, and this was in Oberursel, a suburb of Frankfurt, in the middle of Germany!

Chapter 11
Frankfurt, Germany; July 4, 1948

Early in 1948, the Americans closed down the operation in Oberursel and moved everyone to Wiesbaden, about twenty-five miles away. They didn't lay anyone off but we had to commute to Wiesbaden and back every day. I was one of the drivers, chauffeuring people who worked in the offices, so I still had a job but I was no longer "operating." I'd lost my little grocery and cigarette store in the parts department. When the Americans closed the complex in Oberursel, they also shut down the GI service club so I lost that job too. I was still working for the Americans but the new situation couldn't last. It was winter, the trucks were open and no one liked riding around in the cold, so it wasn't long before they ended the commuting and that was it, I was out of a job.

Working for the Americans, I always got three meals a day, and working in the GI service club at night there was no reason to even think about food ration cards of any kind because I ate and drank for free. My problem now was that besides having no money I had nothing that would entitle me to eat. I could go back to a DP camp, that was your choice, and if I did I'd get a food ration, but I wasn't going to do that again. The only identification papers I had was the booklet with my picture they gave me in Oberursel, but you had to have it stamped every month to prove you were working, otherwise you couldn't get food ration stamps. The stamp was also important because if you weren't working they could assign you to some kind of menial job cleaning up, and they also might question you—if you aren't working, how come you have money? Which I did, because when I lost the job working for the Americans, I started operating on the blackmarket in the *Frankfurter Hauptbahnhof.*

Frankfurt was ten or fifteen miles from Oberursel, a forty minute ride on the streetcar, and the *Hauptbahnhof* was the main railroad station, right in the center of the city. Because there were so many people changing trains and coming and going all day, it became the trade center, the local exchange. The city itself was still a shambles. Frankfurt had been hit hard during the war, people had been bombed out of their homes, and other Germans were coming into the cities, many from the east because they were afraid of the Russians—no one wanted to be in the Russian zone. For many, many people things were very grim. I was part of the chaos too, but for me there was still an excitement. I was seeing different places, meeting new people, and even during this bad era, I could always manipulate to get money. I might go hungry for a day or two, like the time I came home from dancing all night at the casino in the park in Bad Homburg, walking in a pair of shoes that were too tight, and the only thing I had to eat was a can of spinach. It was bitter cold, freezing inside and out, I had no wood to start a fire in the room I was renting, so I opened the can and ate the spinach frozen. But it didn't bother me because I knew the next day I'd be all right. There were still plenty of GI's around and I'd go see someone I knew and ask if he wanted to sell me a carton of cigarettes. They were allowed to buy a carton a week and the GI might say, "I don't have a carton but I'll sell you three packs," because he wanted money to take his fraulein out. The GI would trust me to come back later that afternoon or evening to pay him, and I'd take the streetcar down to the *Frankfurter Hauptbahnhof.* It had become the black market stock exchange where the price was set every day for everything from socks to diamonds—liquor, shoes, stockings, a single cigarette to a carton or a whole case, a pound of coffee to a boxcar of coffee, trucks, tanks, guns, you name it, it was for sale.

People came in there with whatever they owned, they carried dishes, clothing, silverware. If you were going hungry at home, where did you go? Down to the exchange, the *Frankfurter Hauptbahnhof.* I knew the other guys selling cigarettes, so I'd walk in and ask, "What's the price today?"

"Six marks a cigarette."

Some days it would be four marks, some days five or maybe 5.75. It wasn't much. Once I used a five mark note to wipe my ass, just because I had nothing else and wiping my ass was worth more than anything I could buy with five marks. After I got rid of my wares, I took the trolley back to Oberursel, paid off the GI, and then I had money. If a cigarette sold for six, I'd give the GI four-something apiece, so I made about 25% profit. If I sold a whole pack I'd only make five marks on the deal, but if I sold them individually I made more so if I had time I always sold them one by one, even if I came in with three cartons.

The chances of getting caught were very slim. Sometimes they'd have a raid and the German police car we called the White Dream, the *Weisser Traum*, would show up. But we always got word a good half hour beforehand that there was going to be a raid. When the White Dream rolled up the only people in the *Hauptbahnhof* were people with train tickets! Half an hour before there were a thousand people milling around, buying and selling anything you can name—now, it's all emptied out and the only ones left are people going somewhere!

When Albert and I were still buddying around together we spent a lot of time trying to find food because he'd lost his job as a welder when the Americans closed down the shop in Oberursel. One day we traded for a loaf of bread and when we cut into it there was an inch thick space of nothing but air between the crust and the bread. The loaf wasn't bread either, it was a mixture of sawdust and beans. We were so hungry we ate the mass of undigestible matter anyway. It filled us up. Afterwards we were standing in one of the public lavatories they had on the street, having a piss, and Albert said, "You know something, Joe—a full stomach is better than sex!"

While I was still hustling cigarettes, I got a job as a switchboard operator at the *Motorenfabrik*, a diesel engine factory in Oberursel. The job had nothing to do with my knowing any GI's, I got it by applying for it and they hired me because I knew English, as well as German. The company was run by an obnoxious German, although an American officer was in charge because everything was still being run by Americans. There were two girls and myself, each working a shift on the switchboard, so I was free at different times to keep hustling cigarettes for food. Unlike all the jobs I'd had before with the Americans there were no meals, no fringe benefits, and any food I bought I had to buy on the blackmarket because there was still nothing available in stores. There were also fewer GI's around so had I lost many of my contacts for wheeling and dealing.

I was living at a lower level now, moving down the economic ladder, and there was nothing left for Albert in Oberursel or Frankfurt, so he decided to go to Hanover to live with his parents. He had been seeing a girl named Grete who came from the Hohemark area in the Taunus mountains outside Frankfurt. She lived there with her family—her father worked as a forester and was a former German wrestling champion, built like a gorilla—and Grete also had a two year old son. During the war there had been a hospital in Oberursel for German soldiers and while Grete was working there she met a *Ritterkreuzträger*, a soldier who had won the Iron Cross, the same medal worn by the soldier my sisters had met in Maastricht early in the war. While the soldier was recuperating, he and Grete met and fell in love, and she had a child. The war ended and the *Ritterkreuzträger* left, so she was alone. She had been very much in love with Albert but when he left he told me to tell Grete he'd joined the French Foreign Legion. He hadn't told her he was leaving and one day she called me at the *Motorenfabrik* trying to find him. I told her Albert's mother was sick so he'd gone home to Hanover. We arranged to meet in Oberursel at the *Café zur Krone*. She rode down from the Hohemark on the trolley in the afternoon, and we had a cup of bouillon and a piece of roll. We started getting interested in each other, and soon we fell in love. Grete was a beautiful dresser with the most beautiful legs in Oberursel, and because she was so beautiful some of the other Germans were jealous. They didn't dare take it out on me but my dating her caused trouble for Franz.

He was a friend I had met through Albert and because he was German himself he got challenged to fight. He ended up with black eyes and lost teeth, when the guy they really wanted to fight was me!

With all the chaos and uncertainty everyone was living with, holidays were very important, perhaps especially for poor people, and one of the biggest all over Europe was what the Germans called *Fasching,* the carnival just before Lent begins. There was a dance at the *Turnhalle* and they put little tables in the corners where the lovers could sit. Everyone wore a mask so you didn't always know who you were dancing with, then at midnight, you took off the mask and saw your partner. They would also suddenly turn off the lights, then turn them on again, another gimmick to entertain people. Grete was my date and she wore a beautiful Hungarian dress, short like a miniskirt so it showed off her legs. I was so proud of her and very much in love, and we were upstairs on a balcony above the dance hall, sitting next to each other holding hands when the lights went out. We kissed, then the lights came on again and sitting on the bench just a few seats down from us was the *Ritterkreuzträger!* There were rumors that he was in Oberursel but Grete hadn't seen him—and there he was, like a magician had made him appear. The music started again, I took her out to the floor and we started dancing. I was very aware of this man, the father of her child, watching us, and I knew everyone was thinking why is she with Joe instead of the *Ritterkreuzträger?* I knew he was distraught too because of what had happened between them. I held her in my arms and kissed her while we danced, and suddenly—as if she had felt a terrific burst of energy—she hugged me very hard and began crying, crying on my shoulder. She cried all the time we were dancing. I was sorry she was sad, but for me the moment was so beautiful. I had never felt anything like it before, a beautiful woman in my arms, dancing, and holding on to me so hard while she was just crying and crying.

I met the *Ritterkreuzträger* later and I'm sure he was a very courageous man, and every bit as much in love with Grete as I was. He was jealous of me, not personally but only because he loved her—all three of us, me and him and Albert had been in love with her. But he had left. He hadn't wanted to marry her. Then she had his son, and now he had come back.

Several months later, after Grete and I stopped seeing each other, I was walking to the house where I rented a room in Oberursel when I passed a photographer's studio. Coming out the door I saw a girl in a bridal gown and a man in formal wear. It was Grete and him, the man who won the Iron Cross. They were having their picture taken. I was so taken back by seeing them that I didn't respond properly. I just kept walking. I had loved her so much, and when we danced at the carnival everyone cleared the floor, and she cried in my arms. Now when I saw them, I should have gone over and congratulated him, because I was glad she finally married him. But I just kept on walking.

By the summer of 1948, when I was twenty, I was still working on the switchboard at *Motorenfabrik* and going down to the *Frankfurter Hauptbahnhof* every day to sell cigarettes so I could buy food. I was eating but it was day to day. Bread was sold on the street. They would cut off a slice and you bought it by weight, fifty, a hundred grams, and along with a can of Spam I made a sandwich. I had sold the checked jacket I took from the house in Kohlscheid for food and now I was wearing a jacket made out a pup tent. One day in July—by coincidence it was the 4th—around two in the afternoon, I had done my wheeling and dealing at the *Hauptbahnhof* and was walking out of the station when I saw something that got my attention. Frankfurt was still heavily populated with Americans because they had a military headquarters at the I.G. Farben building in the center of town, near the *Hauptbahnhof*, and it was the American holiday so a lot of GI's were out on the street. As I walked away from the train station down the *Frankfurter Hauptstrasse*, I noticed a tall, black GI walking toward me on the sidewalk, a real sharp-looking American soldier with a gorgeous blond on his arm, an absolutely luscious thing. I couldn't take my eyes off her as they walked past, and then beyond, in the space she had just left in my vision, I saw a big GI holding a camera by the strap, bent down with his head tilted sideways to look in a store window. There was something familiar about him and when he stood up and started walking toward me, I saw who it was—Sergeant Hannah! I couldn't believe it. There he was, big as ever, dressed sharply in his OD's, holding the camera by the strap and looking just like a typical GI tourist. When I called out his name it took him a moment to recognize me wearing the pup tent and dyed brown OD pants.

"Little Joe! What the hell're you doing here? You're supposed to be in the States!"

"What're you doing here? I thought you'd gone home!"

"I re-enlisted."

He had a job at the military headquarters in Frankfurt and I told him the story of what happened after he and Captain Hardesty left Wetzlar, how the Kraut-loving captain shanghaied me out of the outfit and shipped me off to a DP camp so I never got into the army officially, and consequently never got to the States.

"Sonofabitch!" Hannah said. "Tell you what—come and see me tomorrow. I'm at the I. G. Farben Building...." He wrote down the phone number and extension. "I'll see what I can do to get you a job with the army again."

We talked for a couple minutes but he had a date, so he took off, and said he'd see me tomorrow.

I couldn't believe it—of all the people to run into. Frankfurt was a big town and we could have both lived there for a hundred years without running into each other. It was like getting picked up by John B. Roper after I dug the foxhole. I felt so good I went down to the *Frankfurter Tiergarten,* the zoo, where they had a band led by Joe Kwikert. He imitated Glenn Miller's sound and just hearing it made your feet want to move, but none of the girls were too eager to dance with a guy wearing a pup tent. The jacket was rather stiff and pretty grimy. I didn't know anything about dry cleaning, although I couldn't have afforded it anyway.

The next day I got up early, jumped on the streetcar and showed up at the I.G. Farben Building. I went to the main gate where they had an MP posted in a booth and told him, "I want to see Sergeant Hannah."

The MP took a look at me and told me to get lost.

"I've got an appointment with him. He told me to call."

The MP didn't believe me.

"Beat it, kid," he told me again.

He was a young punk, impressed with himself in his white scarf, white gloves and belt, and I was wearing the pup tent and dyed pants.

"I've got his phone number," I said.

No way I was leaving—but I couldn't call from outside, the MP had to call through the switchboard for me.

"Just call this number and ask for Sergeant Hannah."

Finally he dialed and spoke to someone. Then he put the phone down and said, "Okay. Stand outside. Stay out of the way."

I went out and stood by the gate they had across the entrance. Five minutes later Hannah came running down from the main building. He signed me in on the MP's clipboard and on the way into the building, he said, "Joe, I've got a good job for you if you want it. You'll be the first civilian driver for an American major-general."

"You're kidding," I said.

Hannah shook his head. "The general wants to see you. I checked it out with him, gave you a good reference, and the job is yours if you want it."

We walked into the office where there was a smart-looking secretary, and Hannah introduced me.

"This is the general's new driver."

She gave me a look. I had long dirty hair with greasy cream on it and was wearing the strange outfit, which probably smelled like hell. But the pup tent and dyed pants were the only clothes I had. At night I folded the pants up and slid them under the mattress so in the morning they'd look "pressed."

The secretary went into the office and a few minutes later came out and said to me and Sergeant Hannah, "The general will see you."

General Dorn was a big, tall man and when he took his first look at me something must have crossed his mind that wasn't very complimentary, but it didn't shake Hannah at all.

"General, I know Joe and he's a good man. I can vouch for him."

"Okay, sergeant," the general said. "Do you have a military driver's license?" he asked me.

"No," I said.

"Will you take care of that, sergeant. I'll see you tomorrow morning, Joe. Pick me up at 8:30. My secretary has the address."

I knew where it was because by coincidence he lived in Oberursel. I left the Farben building and went to the motor pool which was near the *Frankfurter Turm*, the tower. It was administered by the Germans and they were not at all happy to see this guy named Filipovic walk in. I wasn't one of them and somehow I'd gotten the kind of job you had to buy with cigarettes or food or some other commodity. So they questioned me: "Who are you, we've never heard of you. How'd you get the job? You don't even have a license."

"I'm supposed to get one," I said. "Call Sergeant So-and-so."

They reluctantly called the sergeant who of course was American and he was expecting me. They gave me a physical, then a written exam, a color blindness exam, and finally I had the eye test. My vision was pretty good but I knew my left eye wasn't 20-20 and to drive a general you have to have eyes like Ted Williams. A sergeant first class was giving the exam. "Cover your left eye and read the chart," he said. I covered my left eye with my left hand and read everything down to the smallest line. "Okay, now give me the other eye." So I dropped my left hand and put my right hand over the left eye again and read down to the smallest line. I passed and got the license. If I was going to flunk it wasn't going to be because one eye wasn't perfect!

I went out and they showed me the car, a Ford. It had been made after the war but it wasn't brand new. It was a general's car so it had two stars on the front and rear bumper. When the general was in the car the stars were visible, but when he wasn't in it you covered them up with a piece of canvas that slid over them, the same kind of cloth my jacket was made of. The routine was to come in the morning to get the car, then go get a trip-ticket and gas up. I could leave the car at the general's house at night—he lived very close, only a mile and a half from my room—but I had to report in every Friday. They also had facilities where I could wash the car.

The next morning I got the streetcar in Oberursel and arrived at the motor pool at 7:30, got the trip-ticket and gassed up, and drove back to Oberursel to pick up the general. I was on time, he walked out and I did the bit, opened the door for him, closed the door, and drove him to the I.G. Farben Building. When we got there he said, "Joe, I don't need you till 5:00 but my wife wants to go shopping at the WAC Circle in Frankfurt today." The WAC Circle was a shopping complex for Americans, so I went back to his house in Oberursel, picked up Mrs. Dorn and dropped her off while she went shopping. I waited quite a while, she bought groceries and whatever else she needed, then I piled them all in the trunk and drove her back to Oberursel. When I dropped her off she said, "Joe, it's kind of late so I think you should get back. General Dorn leaves the office promptly at five." I decided to take a shortcut on the back roads through a small farming village called Stierstadt, "Bull City," before I hit the *Autobahn*. It would cut off at least five or ten minutes. On the outskirts of the village, just a few miles from the *Autobahn*, I slowed down behind a four-wheeled hay wagon. It was being pulled by one horse harnessed to a long beam. Even though I was in a hurry I was driving very carefully because it was a narrow, secondary road, and it was my first day on the job. The horse was just clopping along and as I pulled out to pass the wagon, I kept my eye on the horse because you never knew when something might scare them. I got past the wagon but as I was passing the horse it suddenly leaped up, there was a CRASH! and a heavy thud on the roof of the car and a beam shot past my shoulder. There was another commotion as the horse fell off the left side of the car, and another THUD! when the driver landed on the roof, then rolled over the other side. By now the horse was running down the road by himself, the reins trailing after him.

"Oh, God!" I moaned and looked around. The car had stalled and lurched to a stop because now it was dragging the wagon. The beam had punctured the trunk and gone right through the back seat, past my shoulder and stopped short of the windshield. I got out and looked around for the farmer. He was lying in the road on the other side of the car and I could hear him groaning "Ohhhhh...Ohhhhh!" I ran around the car. He was stretched out on his back, but nothing was broken, he'd just got the wind knocked out of him. When he got his breath and sat up he started apologizing. "I'm sorry, the horse went crazy and shied!" People had gathered by now and the farmer was okay, although his horse was long gone, and I said, "I've got to get out of here to pick up an American general."

"A general?"

"Yeah, and it's my first day driving for him. I'm sure I'm going to be fired."

The farmer felt bad, he kept apologizing for the accident, and everyone wanted to help me. I got in the car, started it up, and several people braced themselves against the wagon as I slowly pulled forward. There was a nasty sound of scraping metal as I peeled myself off the beam. I got out to look at the damage. The whole trunk lid was caved in like a bulldog's face. The rear window hadn't broken but the roof had four or five deep dents where the horse had kicked it with its hooves, scrambling to try to get off. I looked at my watch and jumped back in the car. Now I had to step on it. How the hell was I going to get out of this mess? On top of it all, I also had to look out for the MP's who patrolled the streets. The two stars on the bumper were covered so they could stop me and ask why I was driving a vehicle in such a condition, and impound it. Nevertheless, I was speeding so I wouldn't be late. At the compound the MP waved me through the gate; the front still looked fine and he didn't see the back as I zipped through and pulled into the general's parking space. I'd made it, it was a few minutes before five, and I sat there wondering what the hell I was going to tell him. The job was gone, that I assumed right away—it didn't matter what Sergeant Hannah told him about me, I wasn't going to be driving any generals around tomorrow!

I looked up and saw General Dorn coming out the front door of the I. G. Farben building. He was walking with his tilted head down, in deep thought, and when I got out and opened the back door for him all I could see was the gaping hole in the back seat, the upholstery ripped, stuffing hanging out and tufts on the floor.

General Dorn walked up and as he was about to step in, I said, "General, I have to tell you something."

He had one foot in the back but I'd caught him just before he saw the inside and he hadn't noticed anything yet.

"What's the matter?" he said.

"I had a little accident."

"Was Mrs. Dorn in the car?"

"No. It happened after I dropped her off at home."

He looked around, saw the hole ripped in the seat, then he walked around the back and saw the punched-in trunk.

"Goddamnit!" he said. "What the hell did you run into?"

He was so tall that unfortunately he had an even better view of the dents on the roof than I did.

"A horse and wagon ran into me," I said, and kept talking. "But I can get it fixed. I know a guy in Oberursel who can repair it."

The general was still shaking his head over the damage but he got in the front seat—obviously he couldn't sit in back—and rode beside me as I drove him home. When we got to his house, I jumped out, ran around the front of the car and opened the door for him. He was already getting out.

"I'm going to take it over and get it repaired," I said.

He didn't say anything, he just nodded like he didn't even want to think about it, and went in the house.

I had no idea how I was going to pay to fix the car but I drove over to a body and fender shop I knew about. It was closed. I drove to the guy's house, found him at home and showed him the car. I told him I had to get it fixed as soon as possible, hopefully tonight, because I was driving for a general.

"Tonight?" the guy said. "This is going to take me four days. I've got to knock the dents out, put on the putty, paint it, it's a big job."

I'm thinking, four days. I'll never get away with this for four days.

"How much can you do tonight?"

He was looking at the mess inside.

"I can sew up the seat and knock out the dents, maybe put a little putty on."

"What's it going to cost?"

He gave me the price: "Five pounds of coffee, five pounds of butter, twenty-five pounds of flour..."

"I don't have that kind of stuff," I said.

"You're driving for an *Ami*, if he wants the car fixed, let him pay for it."

"I just started the job today, he's not going to pay that much to fix it."

The guy shrugged. "What else can I say? That's the price. It's what I'd charge anyone."

I got in the car and drove back to General Dorn's house. I didn't know what kind of reception I'd get, but there was no choice. I knocked on the front door, the maid answered and I said, "I need to see Major-General Dorn." She had a smirk on her face, she had obviously heard about the accident. The general came to the door and I told him I could get the car fixed but it would take four days and I told him the price the guy gave me. His wife was standing behind him and he said to her, "What do you think—do we have that to spare?" She went to the kitchen to check with the cook and while they were looking in the cupboards I kept talking and talking, saying the guy would do a good job, I'd fix everything up—so I wouldn't give the general a chance to tell me to forget it and just wash his hands of the whole affair. His wife came back and said she had the coffee and a pound of butter and she could buy the rest when she went shopping tomorrow. The general wasn't happy but he didn't seem that upset, even when I was proposing to get him involved in a little blackmarketing. When I drove back to the shop with coffee and butter I told the guy I'd bring the balance the next day. But I couldn't leave the car yet because I had to go get a trip-ticket; they issued a new one every day. I only had to go to the motor pool, half way to Frankfurt, and by now it was dark so I wasn't as worried about the MP's. But I had an anxious ride past the guards at the gate and gassed up myself so the GI didn't see the car and report me. When I got back to the shop in Oberursel I asked the guy if he needed any help and he said no, and I told him I'd be back in the morning to pick up the car so I could drive the general to work.

When I came back he had all the dents in the roof knocked out, he'd soldered a piece of sheet metal over the hole in the trunk lid and puttied it up, and mended and patched the back seat. He put the stuffing back in and cut off a piece of some other upholstery and sewed it in. You could see the patch, no question about it, and the red putty all over the trunk and back fenders really stood out against the OD green paint. The guy had done a lot of work in one night but the car looked like a mess and after I drove the general to work, I went back to his house and stayed there until about a quarter after four, then I ran back into Frankfurt. Fortunately Mrs. Dorn had a private car and made a point of using it till the general's was fixed. The body and fender man did as good a job as he could, but even after he painted it you could tell something had happened to the car. The next Friday after the car was finished, I went to the Frankfurt motor pool to turn the car in for the weekend and the guy in charge said, "We've got a new car for you." It was a brand new Plymouth! And that was the last I saw of the Ford with all its patches, inside and out.

Driving for General Dorn was an easy job and he was a decent man to work for, besides the way he treated me when I had the accident. If no one needed me, I could sit around the general's house during the day and read, and I was getting paid good money, 180 marks a month. But the big thing was eating at the general's house. Three meals a day was part of the job and I ate in the kitchen with the rest of the staff. They had two full time gardeners, an upstairs and downstairs maid, and a cook. They lived very graciously with five servants—six with me—working for them. After the way I'd been living when I ran into Sergeant Hannah, hand to mouth every day, it was a very good situation. But it looked as if it wasn't going to last long.

In the fall, General Dorn was promoted to military governor of Baden-Württemburg, a district near Stuttgart, and not only was he moving from Oberursel to Stuttgart, instead of a civilian driver they were assigning a sergeant to drive for him. I was told a new family was coming from the States and most likely they would take over everyone in the household except me. Officially I wasn't part of the staff because I was paid from a different office, and it would be up to the man who replaced General Dorn to decide who drove for him. The man replacing him was a civilian named Mr. Bingham. I was told to pick him up at the Rhein-Main airport and drive him to Mr. Andrews' house. Mr. Andrews was also a prominent civilian, high up in the military government, and he lived about fifteen miles from Oberursel. Mr. Bingham was going to stay with him till the house in Oberursel was repainted. When I got to the Rhein-Main airport, Mr. Andrews was there with his own car and driver but he still had a Ford and when he saw the new Plymouth I was driving he said, "We'll take your car." He sent his German driver home with the Ford. When Mr. Bingham came out of the airport, I just opened the door for him and Mr. Andrews didn't introduce me; they were talking and weren't paying attention to whoever was driving the car. I drove them to a beautiful Tudor style home and when they got out Mr. Andrews told me to wait. The house was a fabulous place and I sat there wondering if this was the last trip I'd make in the Plymouth, if I was about to lose my job. Maybe I'd go find Sergeant Hannah at the I. G. Farben building and see if he could arrange something else. I waited for an hour or so, and when they came out Mr. Bingham said to me, "I understand you were driving for the previous occupant of the house in Oberursel."

"Yes," I said. "Since July 4th."

"Do you drive for anyone else or was this a regular assignment?"

I said, "I only drove for the general, and if you want me to drive for you I'd be glad to. It would be up to you, just for the asking."

"Fine," he said. "You can go home now and come back and pick me up in the morning. Mr. Andrews and I are going in his car tonight."

I thanked him and drove back to Oberursel, thinking I hadn't lost the job yet. When I picked Mr. Bingham up the next morning, he told me he'd be staying at Mr. Andrews' house for two weeks and his wife was arriving next week. "If it's all right with her, we'll keep you as our driver."

And that's what happened. Mrs. Bingham arrived, I drove her over to the house in Oberursel where she met all the staff, and I started working for them. The routine was similar to what I'd done for General Dorn. I drove Mr. Bingham to the office, then drove Mrs. Bingham to wherever she wanted to go during the day. In the evening I often took them to parties. At their high level people were always having dinner or cocktail parties in the grand villas on the outskirts of Frankfurt and when I drove up there might be twenty or thirty cars parked outside. General Lucius Clay was often a guest. He was the supreme commander below Eisenhower for the occupation forces, and I had picked him up several times and brought him to parties at General Dorn's. They had been in the same class at West Point and now they were all part of the military government which was starting to administer the Marshall Plan. Besides shopping or social affairs, I also chauffeured Mrs. Bingham to DP camps. There were still thousands of people in the camps and she took an interest in them, talking to the UNRRA officials, and eventually she became some kind of official herself. She didn't confide in me about what she was doing; she wasn't the kind to brag about things, she just did them, and a couple times a week I drove her to the DP headquarters outside Frankfurt where she was working.

I was on call all the time but I also had a lot of free time so I started doing gardening for them. If I knew I had three or four hours free after I took Mr. Bingham to work, I'd change my clothes and go out and tend to the roses and flower beds. It wasn't in my job description, it was something extra, but I liked it and the Binghams were good to me so it was something I could do for them. Supposedly there were already two gardeners but the cook's husband only worked in the vegetable garden, and the other gardener was by profession a chemist who didn't know anything about flowers and only worked there for the food. The house had a large formal garden which was too much for one man, and since the chemist couldn't make anything grow and was only good at weeding, I planted a lot of flowers. I also had time to read. When I was driving Mrs. Bingham somewhere I often had to wait, so I began taking books in the car and while she was shopping I could read. I had a card for the library in Frankfurt and first I tried philosophy but didn't care for it, so I tried history. I had read *Mein Kampf* when I was in Wetzlar, but I didn't read any history of World War 2 until later, in the fifties. Nor did I read about the First World War which I'd heard so much about from my aunt and the older people. What interested me was ancient history. I was also taking tutoring because I was aware of my ignorance, the fact that I had not had a real education. There was a retired German professor who lived in Oberursel and I had approached him one day when I was working at the *Motorenfabrik* and he said he'd give me two three-hour sessions in exchange for a can of Half and Half pipe tobacco. That was something I could get from the GI's. It was so cheap if I was buying their cigarettes usually they'd just give it to me. The professor taught at the *Frankfurter Technische Hochschule* and he helped me in math and German language, and also recommended books I should read. Sometimes I wouldn't see him for two weeks, but I could call up and tell him I'd finished a book and he'd suggest what to read next. I'd get the book at the Frankfurt library and later he'd quiz me about what I'd read.

As time went on, I got to know the Binghams better. They were from a very old, rich American family—totally different from GI's like Hannah and Roper—but once again I was meeting a kind of person I had never known before. Mrs. Bingham was a Mayflower descendant but she refused to join anything like the Daughters of the American Revolution or the Mayflower Society. Mr. Bingham had a business on Wall Street and had taken the assignment for one year to help administer the occupation in Germany. General Dorn had been a decent man but he and his wife were very reserved and there was something almost untouchable about them. But when I drove Mr. Bingham, unless he had work to do he always sat up front, and every morning and evening we would talk. On weekends they often took their Packard on sightseeing trips and because they took a liking to me they started asking me to come with them. We'd drive around looking at castles and historic and cultural sites, and when we went on a tour I could always translate in case the guide didn't speak English. Mrs. Bingham could speak a little German, she'd studied it in school, but Mr. Bingham could only order a meal and that was his limit. On these trips he did the driving and he always took his two Brownie cameras, one with black and white, the other with color film, to record the sights. Mrs. Bingham would have the cook make up a picnic basket, we would stop in a small village to get something to drink, and then find a nice spot for a picnic.

When we went on our trips the Germans we met were always very gracious, almost subservient, unlike the cook in the house who was a real Nazi and made no effort to hide it. Even after the war she was still a strong Hitler supporter. Her husband was the vegetable gardner and they were refugees from the eastern zone where during the war he had been in charge of a hunting preserve for the high Nazi officals and generals. They had a really good deal working for the Binghams because not only did they eat and get paid, they also had an apartment over the garage. But being treated so well by the Americans did nothing to change their attitude, they were still Nazis. The cook often told Mrs. Bingham that they should have a nice German boy driving for them, and when the Binghams took me on their trips it drove her nuts because not only were they including me on their outing, but she had to prepare the picnic baskets. What really pissed off the cook was when Mrs. Bingham told her to make the kind of sandwiches Joe liked, and also be sure to make a few more because I was such a big eater. My mentality was still that my next meal might be my last one, and if the Binghams ate one sandwich, I ate three. The maids were so embarrassed by the Nazi in the kitchen that one time when she was complaining they said, “We’ll make the sandwiches,” and after that we always had even more food in the basket than I could eat.

The maids couldn’t have been more different from the cook. They were sisters, very decent people who knew I’d been in the labor camp. The one I was especially friendly with was as gentle and kind as a person could be, too simple to be anything but what she was, just a common, hardworking honest person. Her husband had been missing in action on the Russian front and she still had heard no word about him, by now almost four years after the end of the war. The other maid’s husband was a policeman in Oberursel and whenever I saw him he made a point of asking how things were, did I have any problems, did I need anything. Some Germans were friendly that way, and I suppose they meant it. Others might save their arrogance for whenever they thought they could apply it. A few, like the cook, didn't hide what they felt.

The next summer the Bingham's son David came over. He was attending Harvard and when I met him we hit it off right away. He was about my age and we buddied around in Frankfurt. I showed him the places I knew and when his girlfriend Sarah came over, the four of us, with my girlfriend Irina, went dancing at the officer's club where they had big bands playing the latest tunes. Only Allied personnel were allowed but David brought us in as his guests, and we were treated like everyone else.

I had met Irina and her family while I was driving for General Dorn. After I got the job I could afford a little better accommodations, so I moved to a room on Wiederholtstrasse in Oberursel. It was in a very old building owned by a man who had a shop where he made waterwheels. On Saturdays when I wasn't busy I used to help him bend sheet metal to make plates for the waterwheels and we'd have contests to see who could hit the plates harder and faster. The house had a typical European courtyard with a big wooden door for wagons to drive in and my room was directly above the door so there was a wicked draft, and as the weather got colder it was very uncomfortable. The room I let was in an apartment rented by a woman about forty who lived with her older sick husband. Upstairs was another apartment and the German family who moved in after I did had a very attractive daughter, a couple years younger than me named Irina. After I got acquainted with the family they asked me to dinner several times, and when they saw how the woman I rented my room from was always giving me a hard time, they invited me to come and stay with them. I moved in and slept in the living room, then I fell in love with their daughter. Irina was not only attractive, like her parents she was a very sincere and considerate person. By the summer we were engaged. I had no idea what would happen in the chaotic atmosphere, but because of my legal status I couldn't get papers to emigrate and it looked as if there might be no choice but to stay in Germany. I had a decent job with the Americans and instead of knocking around at night with a crazy guy like Albert, I often spent evenings with Irina and her parents. After dinner we talked and played games like charades, and the feeling I had in their apartment was very warm. Living with them I was doing things in a kind of family setting I had never experienced before. That summer was very pleasant, spending time with Irina and her parents, and with David and the Binghams, all of them the kind of people I had not been exposed to.

One morning I woke up and the second finger on my right hand was so swollen I couldn't even close my hand. I had a little prick on the top of the knuckle I'd gotten from the thorn on a rose bush, so I assumed that's where the infection started. When I went to the house to drive Mr. Bingham to work, I told them I had to go see a doctor. Mr. Bingham said he would drive himself and I went down to an office in the center of Oberursel. Every town had a bureau to manage medical treatment, and they gave me a certificate and told me what doctor to go to. The one they sent me to see in Oberursel was a big, blond German with a long *Schmiß* on his cheek, a dueling scar the students all liked to get in college. He looked at my finger and said, "You have an infection, I'll have to operate on it." I took my jacket off, he gave me a local anesthetic and started cutting. He put a drain on it, bandaged my hand with gauze, and sent me home. That night I had a hell of a lot of pain, throbbing so much I couldn't sleep, and by the next morning my hand was discolored and it was excruciating. I called the doctor and he said to come in right away. When he took off the bandage my finger was purple and blue, and you could see two dark lines going up my arm. It was starting to turn black. He looked at it and said, "Gangrene has set in. I'll have to take your hand off."

I couldn't believe what he was saying.

"No," I said. "Bandage me up again."

"You don't have much time," the surgeon said. "I have to amputate or the infection could kill you."

"You're not going to cut my hand off. Just bandage me up again."

I left and walked through town up the hill to the Binghams. I had missed driving Mr. Bingham again and when I got to the house, Mrs. Bingham asked how my hand was.

"Mrs. Bingham, the doctor wants to cut it off."

"Cut it off?" she said. She looked at it and said, "Like hell he will!"

That was not the way she usually spoke.

"Come with me," she said.

She got her coat, backed the Plymouth out of the garage, and drove me into Frankfurt to the 96th General Hospital, an American military facility. She left the car outside in a No Parking zone and when we went in she said to the receptionist, "I want to see Colonel So-and-so. Tell him Mrs. Bingham is here, and it's urgent." He was the commanding officer of the hospital.

The colonel came down almost immediately and greeted her, "Harriet, how are you. What brings you here?"

"This young fellow is my driver and I want you to look at his hand. He had it operated on yesterday and this morning the German doctor told him he wants to amputate it."

The colonel took us upstairs to his office, cut off the bandage with his scissors, looked at my hand and told his nurse to get the corporal. The corporal came in wearing a white smock and the colonel said, "He's got a badly infected finger, so let's give him a shot of penicillin—" he told the corporal how many cc's, "—then you can put some salve on the finger and dress the wound."

"Come with me," the corporal said, and we went in another room. "Drop your pants." I dropped them and he gave me a shot of penicillin in the ass, I pulled up my pants again, he put the salve on the finger, wrapped it all up, and in twenty minutes I was done. I was surprised—was that it? The surgeon wanted to cut off my hand, all this guy did was give me a shot. They also gave me medication to take for the next several days and I was to stay in bed. By now I had a pretty good fever. As we were leaving, Mrs. Bingham chatted with the colonel, thanked him and told him to give her regards to his wife. When we got back in the Plymouth, she said, "Joe, you can stay in the house with us until you get well."

The next day my hand was still swollen and infected but all the purpleness had gone from my arm and it wasn't as painful. I stayed in the house for a week and Mrs. Bingham brought me lunch and dinner, Mr. Bingham came in to chat and to see how I was doing, David stopped by and we talked. When the maids came in to see me they said they had heard of the doctor who had wanted to cut off my hand, and he had been in the *SS*. I figured the guy must have thought he was still at the front. While I was recuperating in the Bingham's house, David took Irina out dancing to a club in Stierstadt. It was another place the four of us had often gone because they always had a good band. His friend Sarah had gone back to the States and it was a joke between us that he was going to take Irina from me. He was tall, dark and handsome, an American, a student at Harvard, and his parents were rich, and he teased me about Irina because she was so dedicated to me. "God, Joe, I can't understand it! What is it about you?" She was a lovely person and he probably was somewhat in love with her himself. So they went dancing and I lay in bed, living the life of Riley while I got well, and it must have driven the Nazi cook out of her mind to see me being attended by the woman of the house. But as I had gotten to know the Binghams, their attitude changed toward me. I was more like a son to them and it wasn't a servant relationship at all.

CHAPTER 12
BUTZBACH, GERMANY; LATE SUMMER 1949

In August, David left to go back to school. I continued the routine of driving Mr. Bingham to work and one morning on the way into Frankfurt he said, "Joe, you could make yourself some extra money if you're interested. I could use you as an interpreter." As part of his job he had to visit various industries in the Ruhr area and occasionally he had gotten stuck because of the language. He knew I could speak German because of the sightseeing trips we'd gone on, and although he certainly could have gotten a better translator than me, someone who was a real professional—or maybe a nice blonde—he was giving me the chance. I said, "Of course, I'd be glad to do it."

The first trip we took was to Cologne, a good hundred miles. We checked into an enormous, beautiful hotel right across the street from the cathedral. Even after all the bombing the Allies had done, it was still there. We arrived late in the evening, too late to get anything to eat, and I went out and walked around. Four years ago, Hannah and Roper and Riley and I had been paddling down the Rhine and running through the streets of Cologne lobbing grenades into buildings and shooting up the streets. That was just some crazy guys raising hell but the bombing had almost totally destroyed it. There was still rubble in places but they were doing a lot of rebuilding, and strolling around looking at things, I was curious to see what they were doing.

We had breakfast the next morning in the dining room with its high ceiling, the waiters were floating all over the place, and right away Mr. Bingham noticed my awkwardness. When I picked up a roll and carefully cut it down the middle with my knife, he said, "Joe, you don't have to do that. Watch me." He took a roll and just broke it open before he buttered it. It would never have dawned on me to break it open with my fingers. My ignorance about how to behave amused Mr. Bingham but there was nothing condescending in it, and whenever I had a question he just showed me what he did in that situation.

We always stayed in good hotels with good food and were often gone three or four nights. Instead of the Plymouth, we took his Packard, and we alternated driving. Usually it was only the two of us; very rarely he brought his secretary. He found it was awkward because when I went with him he got a suite and I stayed in one of the rooms but if he brought his secretary he had to get her a separate room. Once we went to an area we usually visited two plants a day. Riding in the car and having breakfast and dinner we spent a lot of time together and he told me his life story, including his college days at Harvard, and his trips to Peru to visit Machu Picchu in South America. He was proud that his father was the one who discovered the famous Inca site and he described the findings and what they meant. He was also constantly asking me about myself. I told him my background, how I had left Yugoslavia and came to live in Holland, and what the war was like there. He knew I had been in a labor camp and I described how I escaped and joined the American army, unofficially, of course. So we got to know each other quite well.

Occasionally we went to the I.G. Farben chemical plants in Höchst but most of the places we visited were in the Ruhr area, in Cologne, Düsseldorf, and Solingen, which were all in the British zone. We also went to Münster in Westphalia, and to Karlsruhe in the French zone where I.G. Farben had chemical plants along the Rhine. The Marshall Plan was being administered jointly by the Americans, British and French. It had replaced the Morgenthau Plan which had been put in place right after the war and had required the dismantling of German industry. So for the first couple years after the war what little was left of German industry had been dismantled. But the Marshall Plan was different, it was a program to rebuild German industry instead of taking it apart, and now they wound up trying to find out where all the stuff they had dispersed was—what was left of it, what railroad siding it was sitting on—and reconstructing the factories. It was an enormous job. During the war so much had been destroyed by the bombing and Mr. Bingham was visiting these places to get a firsthand account of how the recovery and rebuilding were going.

By 1949 you still saw signs of destruction. At first they had just cleared streets with bulldozers and pushed the rubble up into huge "snow banks", and many times a pile of rubble had a wreath laid on it in memory of whoever was under it. The towns especially were still a mess. Mainz was a shambles. In Cologne you still saw the outside frames and walls of bombed out houses. The progress was fastest in the American zone, then the British, and last the French. The GI's were rich, they ran around buying cuckoo clocks and porcelain to send home, they paid for their frauleins' rooms, and bought them clothes, so the money was flowing. The British and French soldiers didn't have that kind of money so there was a very noticeable difference in the zones. The concept was to rebuild industry first, then homes, and in the industrial parts of the British and French zones they had done a better job of cleaning up and there was a lot of construction. People were settled by now, everyone was living in something, no one was sleeping in the streets. The Marshall Plan also gave Germany loans to buy wheat and sugar and meat so people were beginning to eat better. There was activity and reconstruction everywhere and the feeling you got driving around in all the zones was that Germany was on its way.

We didn't just drop in at the offices and factories Mr. Bingham was inspecting and it was always a very big deal when we arrived. As Mr. Bingham was ushered into the board room to meet with the directors, I stayed with him like a shadow, listening to everything in case he needed a translation. Each factory made different things but the theme was always the same at the meetings. From the chairman of the board it would go down to the other officers, then to the various divisions with each man giving Mr. Bingham a progress report which I would translate. What could their factory produce this year, the next year, what was their potential for the future. In the last quarter they might have used X amount of steel and they had papers showing who it had been shipped to. Solingen bought steel from Krupp to make surgical equipment, ordinary knives, scissors, razor blades, and afterwards inspection teams could go around and check if the warehouse really contained these items. It was a watchdog system. If they produced 50,000 tons of fine steel they had to account for it—they couldn't make machine gun barrels! When we went to the factories, like Kruppwerke in Essen, where they had huge furnaces, they were always very proud to show us how fast they were progressing. "We're functioning at 70% capacity, in another month we'll bring it up to 90%." The tours were led by a high company officer like the president, and the plant manager would join the group to give a run-down of the operation. Those people rarely spoke English so I translated the manager's description of the factory operation for Mr. Bingham, then translated his questions back into German.

I always sat next to Mr. Bingham in the meetings and the first several times, especially, I felt very nervous. Every industry has its own vocabulary so some of the technical terms were new to me, and trying to translate all this information I might have only been functioning on four instead of six cylinders. Also, the Germans can become very verbal using the classic German vocabulary and obviously they were more fluent in this way of speaking than I was. Even if you speak High German, there is a class distinction in the language and several plateaus in the dialect, so it can be almost overbearing. I picked up more of it as I went along and obviously it was in the German officials' interest to make themselves understood. Sometimes I may have been out of my element, but it never bothered Mr. Bingham. The Germans were always very subservient to him because they knew which side the bread was buttered on—if you gave this guy a hard time he could cut your budget by a few million marks—so if they raised an eyebrow about Mr. Bingham's choice of a translator it would have been in private.

At the same time that German industry was being revived and encouraged, there was still a very mistrustful atmosphere. The directors Mr. Bingham was meeting with were not newly planted in German industry. These were the same captains of industry who had supported Hitler—and might have been Nazi Party members themselves—and they were still running these companies now. Some of them might have spent a few weeks or a month in an internment camp after the war but then they were let go and they went right back to their jobs. Nothing had really changed. Had the Morgenthau Plan stayed in effect the ex-Nazis running German industry would have been thrown into camps and left for a few years. But the Americans especially, more so than the British or the French, realized that in order to get German industry moving, they needed to keep these people.

As we drove around and saw all the rebuilding of Germany, it made me very indignant. I wanted to see Germany suffer. I felt like Eisenhower did when he saw the death camps and said, "We're going to make these people pay for this for a hundred years." He was reacting to the moment and later he was more rational, but that emotion still prevailed with me. I was angry and disappointed that the Germans were being treated so well, not just by Mr. Bingham but by the whole American military government. I saw no reason why German industry should be rebuilt or why anyone should be concerned about the Germans getting more food. Looking back now, I can see that Mr. Bingham understood the future better than a twenty-one year old punk who was full of hatred. But it made me upset to deal with these people and see them put back into the same comfortable position they had when they were working for the *Reich*. Just because they had lost the war didn't mean these captains of industry weren't still Nazis. I regarded them as the same as the *NSB*ers in Holland and thought they should have the same mark on them for the rest of their lives as the collaborators, instead of being put back in their jobs. It made me even more angry to think that countries I cared about like Yugoslavia and Poland were not getting the kind of attention the Germans were getting. As much as Mr. Bingham understood and agreed with me, he explained that as a businessman representing his government there wasn't anything else that could be done. These were the facts of life.

I also saw how well the Germans catered to the Americans, how hard they worked to please them, starting from the *Fräuleins* to the domestics and the ones in the motor pools. Now I was seeing the same thing in the upper classes, and I think Mr. Bingham was pretty much taken in. He may have come with a tough attitude toward the Germans—it's easy to wish people ill from a distance—but as he met more and more his attitude might have changed. He met people like the two maids who worked in the house and were such decent people, and others like *Frau* Von Stahl, someone I knew too through the Binghams, who was an incredibly fine person. She was from the German aristocracy and her husband had been killed in the *Wehrmacht*. He had been a professor of philosophy at the University of Magdeburg and under suspicion by the police because he'd been an outspoken anti-Nazi. Meeting people like her, and in the position he had, I'm sure Mr. Bingham's attitude was affected.

Besides going to meetings and inspecting factories, Mr. Bingham was often invited to parties in the evening and he always took me with him. Some of these officials lived in mansions with huge gardens, servants everywhere, the finest cuisine, the best wines and brandies. This class of people never suffered, even in the worst days. For me the anxiety of going into these homes was incredible. I was totally out of my league and very conscious of manners I hadn't been brought up with. But it was the same as when I sat beside Mr. Bingham at the meetings—taking me into these parties didn't bother him. I asked him what I should do when I sat down and saw six spoons on one side and eight knives on the other, four or five glasses in front of me. Which one do you pick up? He said, "Well, either you can watch the hostess and do exactly what she does or if you can't see her well enough, watch me. I grew up in this atmosphere and I always watch the hostess. If she makes a mistake, then out of courtesy to her, you make the same mistake. You don't want to correct the hostess." If anything, my situation amused him, the same as in the restaurants. "As long as you don't tuck your napkin under your chin, you're all right," he said. I had never seen so many utensils. At home you had a fork and a spoon, you didn't need a knife because there was no meat anyway. The anxiety at the parties was like the first few times going into meetings, then it lessened after a while. Many of the people I met were curious about me and asked if I was American because of my command of English. Or they asked where I had learned German. "Where did you learn to speak German so well? Was either of your parents German?" they often asked. Because of my attitude toward the country it was strange that I probably spoke their language better than any other except perhaps Dutch. I had learned High German in school in Holland and could also speak dialects like *Plattdeutsch* which was what the Germans in our part of Holland spoke. "Low German" was what they spoke around Kohlscheid too. I could also speak the Westphalian dialect which was spoken north of the Ruhr area and was considered a better class of German than *Plattdeutsch*. But German was never my private language. First it was Serbo-Croatian and then for a long time I thought in Dutch, and even today if I say a prayer I say it in Dutch. Then it became English.

It was just before I started working as Mr. Bingham's translator that his wife asked me if I wanted to go to the States.

"I'd love to go to the States," I told her. "I've already tried to apply to go any place—Canada, Australia, Venezuela—but they always tell me I can't. The problem is I don't have DP status."

"Why not?"

"They say I don't qualify since I came to Germany voluntarily with the American army."

"I'll see about that," Mrs. Bingham said.

A few days later, after I'd driven her husband to work, we drove down to the UNRRA office in Frankfurt to see the woman in charge of DP's in the American sector. Mrs. Bingham said, "Joe has an unusual situation and I want to find out how we can file the papers so he can get status as a DP. He was captured in Holland during the war by the Germans and taken to a labor camp where he was a prisoner for two years. Then he escaped and made his way back to Holland where he was picked up by the American army, and he served with them as a soldier until the end of the war. Since then he's lived in Germany, but he hasn't been able to be classified as a DP, so he can't emigrate to the States."

I had given Mrs. Bingham the letter Captain Hardesty wrote for me when he originally tried to get me inducted into the army officially, and she showed it to the UNRRA official. She read it and said, "It is an unusual situation, but there is a way to get him status as a DP. If you can get documentation from Holland saying he was a resident of Holland, then that will be sufficient, along with his statement that he was captured and taken to Germany forcibly. We won't ask for anything more than that. The best way to get a document of that kind would be through the Dutch consulate."

Mrs. Bingham said, "Fine, I can take care of that."

We drove back to the house. It was a very hot day so that afternoon David and I went swimming at the tennis club in Oberursel which had been taken over by the Allies. After we swam I lay down on a blanket and had fallen asleep in the sun when all of a sudden I felt someone shaking me. It was David. He said, "Joe, my mother wants you meet somebody. The wife of the Dutch consul is here." Mrs. Bingham had come to the club with us and she was sitting with a woman about her age, obviously a friend, and when she introduced me I recognized that she had a Dutch name. She asked me a few questions, speaking in Dutch—where were you born, when did you first come to Holland, where did you live, what school did you go to?—and I gave her the information. Mrs. Bingham thanked me and said, "That's fine, Joe. We'll see you later," and I went back and went swimming with David. Two days later, after I drove Mr. Bingham to work and came back to the house, Mrs. Bingham said we should drive down to the Dutch Consulate to get my papers. When we got there they had them. Everything was in order. Apparently they had called the town administration in Heerlen and verified the information I'd given Mrs. Bingham's friend. The school had a record of the students who had been enrolled there, and as the son of a guestworker I had been registered with the Dutch government. All the foreign laborers were registered, as well as any member of their family they brought into the country. So they could find out who I was, it was just a question of knowing who to ask and going to the trouble of doing it.

Mrs. Bingham and I went from the consulate to the UNRRA headquarters where she gave the papers to the woman we'd spoken to before, and she said, "Fine, we'll take care of it." Soon after that, I got my status as a DP. Now I was eligible to emigrate, and the Binghams said they would sponsor me. To get in to America you needed a sponsor, unlike Canada or Venezuela or Australia where you signed a two-year contract with the government and agreed to work at whatever job they gave you. It was like working off your bondage and after that you were free to do whatever you wanted. In the States, the sponsor had to put up a bond and guarantee they would help find you a job and support you if you couldn't support yourself. Applying to emigrate, I didn't even have to go to a government office. Mr. Bingham did the paperwork from his office in Oberursel. He asked me a few questions to make sure he had the right information and filed the application himself. A short time later the CID, the Criminal Investigation Division of the army, came around the neighborhood and inquired about me. Another couple weeks later permission came through. I was on the list to go to the States. Obviously none of this would have happened without the Binghams. I had no idea about how to establish a legal identity, and without that it was impossible to emigrate. That was why I had begun thinking I would just have to make the best of it where I was and resign myself to staying in Germany, as much as I wanted to leave it.

The whole experience with the Binghams and the way they treated me was so fantastic and so unexpected. They made me feel this was what America was all about. Their generosity was like the first impression I had of GI's when they liberated Heerlen, tossing candy and cigarettes to everyone, bringing cans of food into Leo's house. Of course everyone wasn't like the Binghams, or like Hannah and Roper either, but what I saw in them, and what they were willing to do for me—the whole

Bingham family, including David who treated me like a friend, as if I was on the same level as him—all this was incredible, it was something I never could have imagined. Because I had never met people who behaved like this before.

In October the Binghams left Oberursel and went back to the States. Now I was just waiting for notice to go to Butzbach, a hub where you went before being sent to a port for embarkation. The Binghams had assured me everything was in order and I wasn't worried things would go wrong, the way it had after the end of the war when the Kraut-loving captain who replaced Hardesty threw me out of the army. I knew they had to ship out a certain number of people before they brought in any new ones and it was just a matter of going down the roster until they got to my name. Mrs. Bingham had asked *Frau* Von Stahl and her boyfriend, a former American navy officer who was now working in the CID, to make sure I wasn't in any kind of need until I got notice to come to Butzbach, and he stopped by almost every day in his little Volkswagen with food and cigarettes. I had enough money because I'd saved quite a bit during the time I'd spent driving for Mr. Bingham. I'd eaten meals at their house and didn't pay Irina's family much for the room in their apartment. I had friends who looked out for me too. At the *Turnhalle* in Oberursel I was sort of a local celebrity and the owner always asked me to sit with him and his wife at the *Stammtisch*, the regulars' table, and he wouldn't let me pay for food or dancing. So the period of waiting was almost like a holiday. I didn't worry about spending the money I had because in the papers saying I was accepted to immigrate, they said you couldn't bring any foreign currency into the States. I'd arrive broke, but the Binghams would be waiting for me when I got there.

It was about three weeks before the letter came. I said goodbye to Irina and her family, and what I felt leaving them was something new for me too because I'd never really been part of a family before, which was how they treated me. Irina and I were still wearing the rings we had given each other; the custom was to wear it on your right ring finger until you were married, then you changed hands. Neither of us knew what would happen but once I had a chance to leave Germany for the States there was no question.

Butzbach was about twenty-five miles away and *Frau* von Stahl and her boyfriend picked me up in his car and gave me a ride. Everything I had was packed in a big, custom-made leather suitcase, a monstrosity with a belt wrapped around it. The camp at Butzbach was another former German army barracks consisting of two-story brick buildings, and the beds were all double-deckers jammed so closely together you could barely squeeze between them. The rooms for families weren't quite as congested as the ones for single men. They had everyone jammed in because there were thousands of us in the camp, but the mood was very jovial. Everyone was so happy it was like we were high. All of a sudden—after what everyone had been through—we were finally on our way. For most of the people it was the first time in their lives they'd had a future.

The movement of people in and out of the camp was constant because sometimes three shipments a day left to go to the ports of embarkation. There were so many of us in the camp you hardly noticed they were gone. Then more people came in, got their sheets and blankets, and moved in. There was nothing to do all day but mill around and check the lists. Every day they posted the new list of names on a long board, and beside the name was what ship you were assigned to and the port it left from. If you were going to Australia, you were sent to Naples, to Le Havre if you were going to South America, and to Bremerhaven if you were going to America. Those were the three main points of embarkation. You went to check the lists every day and there might be several hundred names on the list with everyone jostling to see the board so you had to fight your way in to look. It would take me half a day just to go through all the names because like everyone else I checked every sheet, not just the "F"s. When you didn't find your name among the F's, you went to the A's, and so on, looking at every single name on the board to find yours, just in case they had put it in the wrong place. Our occupation was waiting and looking at lists, but people were too excited to be bored. They kept talking to everyone—"Where're you going? Are you going to the States? Who's your sponsor?" You could always tell the people whose names had appeared on the list to leave the next day because they were the ones who were the most excited.

We had all come together and met each other in Butzbach from all over Europe—Russians, Poles, Jews, Yugoslavs, Latvians and Estonians, Czechs and Hungarians, some Romanians and Bulgarians. Whenever I met a Yugoslav I always asked if they'd come from anywhere near Vrbosko but I never met anyone who did. Hardly anyone spoke about what they had been through, now they were just so relieved to be living. Someone might ask, "Where were you?" and I would say, "I was in the coal mines." Someone might say, "I was in Essen." Or "I was in Münster." That was enough. People didn't elaborate much. The only person I met who talked about what he'd been through was a Jewish fellow who joked about his experience when he showed us his body. I was in a room with all Jewish guys, the only Christian. When you first arrived, you went up to the attic because all the rooms in the first two storeys were filled, then as people left you "graduated." You grabbed your bedding and ran downstairs and staked out a bed. The authorities didn't care—if they'd come in and tried to move us we would have thrown them out the window! When I ran down I was the first one in the room. I took a top bunk against the wall and tossed my blanket and sheet on the bed. I no sooner had my stuff on the bed than another horde came in with bedding on their shoulders, eleven of them, jabbering in Yiddish. I spoke to them in German and we introduced ourselves. They were all going to the States and they were all young except for the one guy who'd been chopped up so much you couldn't tell how old he was. He wasn't even five feet tall, a skeleton, maybe eighty pounds. He was wearing a cap and a suit, and constantly moving inside his suit because it was so big for him. He was a very happy guy, especially for someone who had gone through such hell. We were all sitting on our bunks with our feet hanging down when one of the guys mentioned a camp he'd been in and told us about his experience. The little fella picked it up and said, "Well, what happened to me was, they left me for dead. They had done all these experiments on me, cutting me open to see if they could put me back together again. One day they must have thought I hadn't made it because they put the sheet over me and shoved me off to the death house. Then I woke up. I sat up and saw all these bodies around me covered with white sheets—and I had crawled out from under a sheet myself! I was sitting there looking at all these bodies when a nurse came in and he said, 'What're you doing sitting up?' I said, 'What am I doing *here?'* The nurse said, "You're supposed to be dead, that's why you're here!' I said, 'Well, I'm *not* dead!' The nurse ran out and a few minutes later a couple doctors came in to see for themselves, and when they saw me they couldn't believe it. So they rolled me out and put me back in the ward again."

This little guy just refused to die. He was crisscrossed with stitches from some kind of medical experiment, scars all over his chest and stomach and all the way around his back. The other guys in the room had all been in labor camps but this guy had been in some "research center" where they were going to just cut him up till he died, then send him to the chimney.

He was the one who introduced me to the system of signaling the Jewish guys had after he walked in on me interrupting my "physical needs," so to speak. Afterwards, he said, "Joe, we've got to get you a key." If you had a woman in the room, you put the key in the lock and turned it half way so no one could put their key in from the other side. That way they got the message it's taboo, don't come in. If a guy was outside with his girl, he'd have to wait. Or he could say, "Let's try your place."

Everyone knew they were just passing through Butzbach, it was temporary, but we still had to eat and the food in the mess hall was awful—watered down hash, dehydrated potatoes, powdered eggs. The reason it was so bad was because the cooks were selling all the food on the blackmarket, so we got the dregs. By now we were accustomed to a little better food, especially myself. For more than a year I'd been eating with the Binghams and the Dorns and my eating habits had changed, I'd become more selective. In Butzbach the only time I got good food was when I was invited to someone's room or when someone brought me food and then I shared it around with the others too. Irina's father came from Oberursel to find me and he brought food and sandwiches, and *Frau* von Stahl and her boyfriend came over more than once with food and then I had plenty to share with the others. People were also selling things, usually for food—we were buying the food the cooks were supposed to be serving us!—and after I spent the few hundred marks I'd brought with me I started selling my clothes. I sold a double-breasted brown suit with white stripes, and whatever it was made of—it might have been bark—when I got caught in the rain after I walked a girl home from a dance, the pants shrank about four inches. I sold that one to a short guy!

There were so many people moving through the camp it was a surprise to meet someone I knew, but one day checking the lists I bumped into a guy who recognized me. Sasha was a tall skinny Polish guy who'd been one of the DP's on the truck when I was hauling wood at Wetzlar, and he remembered me because I had taught him to drive. He told Sonya, the Polish girl I met in Butzbach, that I was the best driver in the world! Sonya was there with her family and they had quite a bit of money with them because they were going to Australia and weren't as concerned about dumping it as I had been, so they bought food and always had plenty which they shared with me. Sonya's family had been deported from Poland to work in a German factory during the war, and Sonya had worked in someone's household as a maid. Her two brothers were missing, no one had seen them since the war. The family had made an inquiry with the Red Cross but the Red Cross hadn't found them. A lot of people still hadn't found their families, or they had just disappeared. Sonya's family left Butzbach without hearing anything, although if the Red Cross located her brothers they could put them in touch, like they did when Albert found his mother and father. Sonya's father was a kind man and had been through a lot but he wasn't a broken man. Her mother was rather quiet, and Sonya was too. She had gone through some hell but didn't talk much about it. She was a beautiful blonde, blue-eyed peasant girl, and she was big—bigger than I was, at least twice! When they left for Australia I wanted to go with them. Obviously that made no sense, it was crazy—I'd just left Irina to go to the States. The States was where everyone wanted to go and the Binghams would be waiting for me when I got there. But in spite of that I would have given it all up and gone to Australia with Sonya if I could. I even went to the authorities and tried to change my destination. It was too late by then, but I meant it. I would have done it.

To understand what it was like to be Josko—or Sonya, or Sasha, or any of the people in a place like Butzbach—you would have to understand how we felt about what we had all lived through. The first feeling we had was to get *away* from this place, to get out of Germany. The country weighed you down, it was like something heavy hanging around your neck, an albatross. For the people who had been in the labor camps, working in coal mines and factories and farms for the Germans, treated as if we were the *Untermenschen,* always reminded we were people who had no right to live a life, now suddenly we had something, and we wanted all the things we'd been deprived of. For me that was, more than anything, love, not just with a woman but the kind of love I never had as a boy. When I met Sonya she was so beautiful and sensitive, and she and her family were also the same kind of people as me. They had been pushed and abused, they had lived in a camp and been through hell too, and now all of us were like the core of the world fanning out to everywhere, to Australia, to New Zealand, to Venezuela, to America. When Sonya and her family left and I couldn't go with them, I felt torn apart at the leaving. I knew I would never see them again.

I had been in Butzbach almost four weeks when the day came that I saw what I was waiting for—my name was on the list to leave for a ship. The next morning I boarded a train for Bremerhaven. Everyone felt the excitement that we were finally leaving, but what I also felt was my hunger. All the people who'd been sharing their food with me had already left, and I had run out of things to sell for food. I had even sold the engagement ring Irina had given me. I meant no disrespect to her by trading her ring for food, but I had so little to eat once the people I befriended were gone there was no choice. The trip took most of the day, almost three hundred miles, and they had told us not to bring food on the train but in the compartment with me was a Jewish family and they had more food than I'd seen in one place since eating at the Binghams. I sat there watching them eat with my stomach making a loud noise and when they finished they took the scraps, carefully wrapped them up and threw them out the window. They didn't even offer me the scraps!

In Bremerhaven the train pulled up by the edge of the water and stopped by the pier. When we got out they lined us all up in front of the train. We could see the ship, the *G.M. Muir*, with a gangplank running up to it and people hustling around, American military personnel going back and forth, talking to the officials in charge of the trainload of people. There were 1,100 of us and we all had a tag with a number on it, pinned on the left side of the shirt or coat. After they lined us up according to the numbers, they herded us into the various compartments, women in one section, families in another, and single men in a third. I was standing in line waiting to board when the guy in front of me turned around and asked me in German, "What are you?"

He was about my age, and I said, "I'm a Yugoslav."

"So am I. Where are you going?"

"Montclair, New Jersey," I said.

"Where's that?" he asked.

"Somewhere near New York. What about you—where're you going?"

"I'm going to Montana."

"You have family there?" I asked. I had never even heard of Montana.

"No, a church group is sponsoring me."

We didn't know where these places were that we were going to.

We started up the gangplank past the sergeants and privates; it was an army troop transport ship so they were all army personnel. When we got below deck a sergeant yelled at us:

"Take whatever empty rack you see, put your stuff on it, and don't move around till everyone has a rack, otherwise you might find you're sleeping with someone else."

The racks were hung up on chains; you let the chain down, climbed in and slept on the rack. I put my suitcase on a rack one below the top and said to my friend, "God, I'm hungry!" I was in pain from not eating.

"I've got some bread," he said.

"You're not supposed to have food," I said.

He opened his knapsack and showed me a hunk of bread, about half a loaf. He broke a piece off and I ate it.

"Didn't you have any food at all?" he said, watching me eat.

"No," I said. "I'm starving. I've been on the train since this morning."

A few minutes later a sergeant first class came in and yelled out, "Who speaks English here?"

I sort of looked at him before I said anything, waiting to see what he wanted someone for.

"I need an interpreter for the galley—down in the kitchen."

When I heard that, I said, "I speak English."

"How well do you speak it?" he asked me.

"I speak it well enough."

"How about Polish—you speak Polish?" he asked.

"Yeah, I speak Polish," I said.

"Well, I need an interpreter down there," he said.

They rotated the passengers for duty in the kitchen, peeling potatoes, cleaning up, helping the cooks, and there were a lot of Poles on the ship.

"Go down below," the sergeant told me.

"Where is it?"

"Just keep going down, you'll find it. Report to Sergeant Jones."

I went down, kept going down the narrow metal stairs until I wound up in the kitchen where all the guys were black, all except one little Frenchman cutting up meat. They were all dressed in white too, the cooks' uniforms, with hats, standing over big pots, stirring and cooking. I asked for Sergeant Jones and a big black guy standing on a short step ladder said, "I'm Jones, what do you want?"

"The sergeant sent me down to be your interpreter."

He looked at me and said, "You look pretty hungry."

"Am I ever hungry!" I said.

He got down from the ladder, filled a plate from a pot, handed it to me and said, "Here. Sit down and eat."

The other guys in the kitchen were all watching, smiling, saying hello, how're you doing?

When I finished eating the sergeant told me after they served dinner that night some passengers would be sent down to wash pots and clean up and he'd give me instructions to translate for them. I didn't have to help, my job was translating. All the single passengers had to work in the kitchen or scrubbing hallways, sweeping the movie room, nothing heavy, just helping out. There were enough Jews on board to cook separately and the kosher kitchen was right next to us through a hatchway. The rabbi came in every morning and said what he had to say over the pots and pans. It was always a big event but they didn't need an interpreter and the Jewish passengers worked in the kosher kitchen. Apparently their cooks weren't very good because they all came over and ate with us! Doing the interpreting I struck up a friendship with some of the crew and they invited me to eat in their kitchen which was where the cooks ate too. So I got plenty to eat and I also started operating, buying cigarettes from the guys on the crew who always had extra, and selling to the passengers.

We had been out at sea several days when we hit a bad storm. People were sick all over the place. Fortunately it never affected me, and since I could speak several languages they put me in charge of the seasickness pills. I went from compartment to compartment to give them to the people too ill to get up. Some of them offered to buy extra pills—and that's when I realized that everybody had money! I was the only dope on board the whole ship! What were they going to do to you? The worst they could do was take it away from you—they weren't going to throw you overboard! So people were trading, buying and selling like crazy, but I didn't sell any pills. No one counted them, they just handed them to me, but I didn't want to get in trouble. If someone couldn't get up from their bed, I just gave them three or four pills and said I'd be back tomorrow. A real doctor! The storm lasted for about three days and it was so bad no one was allowed on the deck because the waves were washing over it. My Yugoslav friend spent most of the time on his back he was so sick. The ship heaved so much some people got hurt; they fell down the stairs, broke arms and legs and collarbones. The ship's doctor was pretty busy. Lying in my rack I could hear the metal creaking—CRAAAACK!—when a wave hit the ship. It sounded like all the rivets were popping loose, and I lay there praying the tub made it through.

The storm put a real damper on things but when it calmed down again everyone's spirits came back right away. By the time the storm was over I was known all over the ship. One of the sergeants had given me an armband to wear for identification when I was distributing pills and I was known as the guy who could get things for you. "If you want butter, chocolate, cigarettes, ask the Yugoslav guy with the armband." All the sailors had something to sell and people were still paying blackmarket prices for cigarettes, the same as they would have paid at Butzbach or the *Frankfurter Hauptbahnhof.* If people had scrip money, the sailors could spend it in Europe, and many of the refugees had dollars too. I didn't try to make money, I was just accommodating the Americans, moving things between them and the passengers. So the rest of the trip was pleasant and I spent some time with a Yugoslavian girl going to Boston to live with her uncle who was sponsoring her. She had lost everyone; her parents, her sisters and brothers were all dead, and the only one she had left was her uncle.

Crossing the Atlantic usually took ten days, but it took us eleven because of the storm, and we were arriving at Christmas so not only was there the excitement of landing but there was all the preparation for a party. They set up a Christmas tree in the movie room, women were busy wrapping gifts, and everything was very festive. There were a lot of children on board and the authorities must have provided the toys for them because there were gifts for all of them. The party for the children was first, people sang carols, and then they had a Midnight Mass. There was a priest, a chaplain, and plenty of rabbis. It was Christmas eve and the next day we were going to land in New York!

The next morning I got up early and people were already up on deck to be the first to see something. When I went down to the kitchen the cooks said, “We’ll hit land by one o’clock this afternoon.” I was still in the kitchen after breakfast when all of a sudden word came down, “We can see mountains!” I ran up the stairs to see. It was a drizzly, cloudy day, and way in the distance I could see the “mountains,” the outline of hills and shapes on the shore. I thought the ship was going to tip over because everyone was up on deck, all crowded over on one side trying to see land, while the ship slowly plowed on through the water. Finally, around 3:30 in the afternoon, we saw the harbor and the Statue of Liberty, and people started getting really excited. Then the boat suddenly slowed down. It came to a halt, the anchor went out. We weren’t moving. We all looked at each other. “What’s happening? What’s wrong?” Over the public address system there was a whistling to get everyone’s attention and then the captain announced that New York harbor was closed due to the Christmas holiday, so we were going to have to spend the night on the ship.

Right away people were worried. ”Oh, my God, they’re going to be waiting for me.” But the authorities had anticipated the delay and they announced people had been notified by telegram that we wouldn’t be debarking until the next day. Everyone had already packed their suitcases and we’d had a final detail to clean everything up. This was supposed to be our last day on the ship and we were aching to land. There was a huge, momentary disappointment, but it didn’t last long. Then the mood changed, people began to get excited and start talking again. It was Christmas day, the harbor was closed, and the night before we’d had a pleasant Christmas eve. After what everyone had been through, what was one more day of waiting.

As it began to get dark, we saw lights appearing—the southern tip of Manhattan and off to the other side the Statue of Liberty. Everything was lit up. No one had ever seen so many lights and it was just like when we first sighted land, everyone was up on deck. We were all leaning toward land, trying to work our way over to the railing to get as close as we could, leaning and looking at the city. People kept saying, “Look at the buildings! Look how tall everything is! Look at the lights!" We couldn't imagine anything so big. All these massive, tall buildings and everything blazing with lights.

No one wanted to go to sleep. It had been more than twelve hours since we'd spotted land but there were still hundreds of people on deck, just out there, freezing their chops off. It was four o'clock in the morning before I finally went down. It was cold but the anticipation was so high, and the urge to get on land was something incredible.

Chapter 13
Boston, 1975

"Stefan?" I said to Joe. "Your father's name was Stefan?"

It was late one night, we were nearly done recording his story, and I realized I had never asked Joe what his father's name was.

"Yeah," Joe said. "Stefan Merkas."

"That's your *son's* name!" I said, stunned to hear this. "You gave him your father's name. Was there any significance in that?"

"No. It was just...my wife felt it was a nice name. She named him Stefan Josef Filipowic."

"Didn't you have any feeling about giving your son your father's name?"

It was such an obvious effort by his wife to join the two men, Joe and his father, through his son, that Joe must have noticed.

"No, I just thought it was a good name. It had nothing to do with my father—I wish the hell it had."

Joe had enjoyed telling much of his story, especially the events after he left home and Josko became Little Joe. But forcing himself to go back to his early years at home had been a painful and wrenching ordeal. Those sessions were all a repeat of his struggle to begin that first day we sat down with a tape recorder between us. At some point during those recollections, as the pauses became longer and longer, Joe would finally shake his head, apologize, and say he couldn't go on. The tap had shut itself off. He could not force it open.

It was not just the memories themselves; Joe also worried that he was creating a false portrait. He kept saying his story was really a comedy, that things like stealing sandwiches from the German coal miners and being picked up by the American GI's were the kind of events that expressed who he really was. Not the stories about his life at home.

"I don't want people to think this is a sad story," Joe said more than once. "Because to me, that's not how I think of it myself."

It was what he'd said the first day we began recording, denying the sadness in his story.

"It doesn't end sadly at all, but some of it is very sad, Joe," I said quietly.

"It's *not* a sad story," he repeated emphatically. "The reason it's not is—whatever happened, I'm here. I made it."

"Yeah, you made it," I said. "But your relationship with your father, the way you were treated at home—"

"Okay," Joe interrupted. "I'm almost inclined to say, 'Yeah, you're right.' Because *that* is part of why I was able to survive."

Joe was lighting a cigarette. He exhaled and gathered himself.

"See if this makes sense, Bill. You know, I've been obviously—I've been giving it a lot of thought. Why does a guy on Christmas day look out a steel-barred window into nothing and everyone around him is crying—just desperate crying, crying, crying, in a barracks filled with sick and dead and dying men, and think—'This is a *relief!'* This has bothered me a lot."

His voice had become intense, almost fierce, as he took himself back to the labor camp.

"You have to see the scene, Bill. We had worked all day in the coal mines, just like any other day, then walked three kilometers back to camp, stood formation, got counted off, and went to the barracks. The barracks were one story, built on concrete with windows about fifty centimeters high running from one end to the other at ground level. They opened with a sliding wooden frame and I lay there on my stomach with my head by the open window. I was breathing in the cold, fresh air because it stank so much in the barracks, and for someone like me who was stealing food and surviving, it was intolerable. The stench from sickness and death was tolerable to those who were sick themselves. They weren't aware of it. Others were on the fringe, they were still sane, though they knew they were sick. You went in stages of dying. When you came in you were strong and healthy. After a few weeks you didn't feel well. Next you felt pretty damned sick, then you became very sick. At the next stage you weren't aware of anything. You would die without knowing it, or if you happened to live till Friday when the truck came, they'd take you away. Men were dead in there that night. Stiff. There were people somewhere who cared about those dead men.

"For me—I was a kid, fifteen years old—but most of the other men in the barracks were older, some of them as old as I am now. Some of them were weeping alone, thinking of their families, their parents and wives and children, good memories. It was the saddest day of the year for them. It was dark and the lights were out because of the bombing but from the houses outside the camp I could hear singing. Some Germans who lived in those houses were singing Christmas carols. But inside the barracks there was no singing, there were no ceremonies, no prayers, no one said, 'Let's have a Mass.' Lying there on the floor looking out through the window, with the others crying around me and I wasn't sad, I knew I was different. I wasn't missing anyone. I wasn't tied to anyone, no one was tied to me. But when I think back now, many years later, on that Christmas in the camp, I think, 'What a hell of a way to grow up.' It wasn't right. Yet, if it hadn't been for that, I don't think I would have made it. I never would have been so belligerent, I never would have been thinking as quick as I did. Because as long as I wasn't home, I could stand almost anything. To me, at that age, it was like an adventure. Had I had a decent home, there would never have been an adventure."

Afterword

The "adventure" Joe began telling me more than forty years ago ended in New York harbor on Christmas 1949. What happened next when he stepped ashore is another story, with goals more ambiguous and elusive than escaping his father, the Nazis, and the coal mines of Heerlen.

Soon after Joe landed and filed his papers to become an American citizen, he was drafted. The twenty-three year old veteran of World War 2 was inducted into the army officially and sent to Korea where he served as a combat photographer. Back home in New York, Joe used his training in photography to get into the motion picture business. He worked on industrial films and in advertising, and eventually set up his own editing shop on 45th St. in Manhattan. With its supercharged can-do attitude and rich broth of humanity, the city suited Joe perfectly. So did the world of film and television where his improvisatory skills and ability to work under pressure helped him thrive.

What interested Joe most were programs about the real world, documentaries and public affairs programs. He took a job with CBS News and moved into editing longer documentaries, working with some of the most esteemed television reporters and producers of the time. He was hired by an experimental public television program affiliated with WNET and quickly established himself as the savior of inchoate documentaries, the editor who gave shape to the miles of film footage shot out in the real world. It was a heady time in documentary filmmaking, spawned by the idealism of the 1960s and made possible by the new portable 16mm film technology. Joe was right in the midst of it, at the peak of his professional life. But soon the ferment and the opportunities began winding down.

The subsequent years became a struggle. The challenging projects came less frequently and Joe chafed at some of the pedestrian productions he was offered to work on. His marriage collapsed and his relationship with his two children was seriously strained at times. There was more than one long separation. But the answer to a question that any sensitive reader will ask is: No, Joe did not abuse his children the way he had been abused. Their lives were not immune to the effects of the early years on their father's life, and he let them down at times. But he loved them, and was haunted by the fear that when he failed them, "they would think the same of me as I think of my father." They didn't, but that fear was part of the psychological burden which Joe carried through his adult life, a burden which became heavier as the years went on.

In his seventies, Joe was diagnosed with cancer, not of the lung despite years of heavy smoking, but multiple myeloma. He was treated successfully, and for his son Stefan the last five years with his father were a chance for them to get to know and appreciate each other more deeply. For his daughter Heidi it was harder to reconcile, but she did. Joe even found it possible to forgive his own father for what he had done.

Joe often seemed to me like he was history itself—the son who inherited the punishment for his father's sins, a child of guestworkers and war who survived some of the worst crimes of the 20th century—although, as he would always remind you, millions of others had had it worse, far worse. Then he was part of the great migration of homeless people fanning out across the globe with their hopes and their energy. What carried him through his adventure, as he called it, was a characteristic everyone noticed in him, his terrific will to live. He expressed this spirit in its most humble form one night when he went back again in memory, very reluctantly, to describe what he felt during the execution in the labor camp, that cold March day unloading logs when others were shot beside him:

"That they could do this...that they could take this away from you, and you could do nothing..."

He hesitated, breathing heavily. There weren't any words for this experience.

"...because what else do you have? I mean, what is your trade?"

"Your trade?"

"Your trade, your..."

Another long pause, then to explain what he meant Joe asked himself a question, something I'd never asked him:

"What is it you love more than anything?"

He thought for a moment and then gave his answer:

"You love life...the only thing you could love more would be a wife, or a child."

With his daughter Heidi beside him, Josef Filipovic died peacefully at Brigham and Women's Hospital in Boston, on December 23, 2012. He was eighty-five.

www.ingramcontent.com/pod-product-compliance
Lightning Source LLC
La Vergne TN
LVHW051500080426
835509LV00017B/1844

* 9 7 8 1 6 2 1 3 7 3 4 4 5 *